Innovations of

KANSEI ENGINEERING

Industrial Innovation Series

Series Editor

Adedeji B. Badiru

Department of Systems and Engineering Management
Air Force Institute of Technology (AFIT) – Dayton, Ohio

PUBLISHED TITLES

Computational Economic Analysis for Engineering and Industry
Adedeji B. Badiru & Olufemi A. Omitaomu

Conveyors: Applications, Selection, and Integration
Patrick M. McGuire

Global Engineering: Design, Decision Making, and Communication
Carlos Acosta, V. Jorge Leon, Charles Conrad, and Cesar O. Malave

Handbook of Industrial Engineering Equations, Formulas, and Calculations
Adedeji B. Badiru & Olufemi A. Omitaomu

Handbook of Industrial and Systems Engineering
Adedeji B. Badiru

Handbook of Military Industrial Engineering
Adedeji B.Badiru & Marlin U. Thomas

Industrial Project Management: Concepts, Tools, and Techniques
Adedeji B. Badiru, Abidemi Badiru, and Adetokunboh Badiru

Inventory Management: Non-Classical Views
Mohamad Y. Jaber

Kansei Engineering - 2 volume set

- Innovations of Kansei Engineering
  Mitsuo Nagamachi & Anitawati Mohd Lokman
- Kansei/Affective Engineering
  Mitsuo Nagamachi

Knowledge Discovery from Sensor Data
Auroop R. Ganguly, João Gama, Olufemi A. Omitaomu, Mohamed Medhat Gaber, and Ranga Raju Vatsavai

Moving from Project Management to Project Leadership: A Practical Guide to Leading Groups
R. Camper Bull

Social Responsibility: Failure Mode Effects and Analysis
Holly Alison Duckworth & Rosemond Ann Moore

STEP Project Management: Guide for Science, Technology, and Engineering Projects
Adedeji B. Badiru

Systems Thinking: Coping with 21st Century Problems
John Turner Boardman & Brian J. Sauser

Techonomics: The Theory of Industrial Evolution
H. Lee Martin

Triple C Model of Project Management: Communication, Cooperation, Coordination
Adedeji B. Badiru

FORTHCOMING TITLES

Essentials of Engineering Leadership and Innovation
Pamela McCauley-Bush & Lesia L. Crumpton-Young

Industrial Control Systems: Mathematical and Statistical Models and Techniques
Adedeji B. Badiru, Oye Ibidapo-Obe, & Babatunde J. Ayeni

Learning Curves: Theory, Models, and Applications
Mohamad Y. Jaber

Modern Construction: Productive and Lean Practices
Lincoln Harding Forbes

Project Management: Systems, Principles, and Applications
Adedeji B. Badiru

Statistical Techniques for Project Control
Adedeji B. Badiru

Technology Transfer and Commercialization of Environmental Remediation Technology
Mark N. Goltz

Innovations of KANSEI ENGINEERING

MITSUO NAGAMACHI
ANITAWATI MOHD LOKMAN

CRC Press is an imprint of the
Taylor & Francis Group, an **informa** business

Published by JSA-Japanese Standards Association, Tokyo, in 2003, in Japanese.

CRC Press
Taylor & Francis Group
6000 Broken Sound Parkway NW, Suite 300
Boca Raton, FL 33487-2742

International Standard Book Number: 978-1-4398-1866-4 (Paperback)

Library of Congress Cataloging-in-Publication Data

Nagamachi, Mitsuo, 1936-
Innovations of kansei engineering / authors, Mitsuo Nagamachi, Anitawati Mohd Lokman.
p. cm. -- (Industrial innovation series. Kansei engineering)
Includes index.
ISBN 978-1-4398-1866-4 (pbk. : alk. paper)
1. New Products--Design and construction. 2. Design--Human factors. 3. Human engineering. 4. Production engineering. 5. System design--Psychological aspects. I. Lokman, Anitawati Mohd. II. Title.

TS171.4.N34 2011
658.5'75--dc22 2010030402

Visit the Taylor & Francis Web site at
http://www.taylorandfrancis.com

and the CRC Press Web site at
http://www.crcpress.com

Contents

Preface xi
About the Authors xiii

1. The Emergence of Kansei Engineering 1
- 1.1 Developing a Product That Is Valuable for Consumers 1
- 1.2 The Emergence of Kansei Engineering 2
- 1.3 What Is Kansei? 4
- 1.4 Kansei Is Something Comprehensive 5
- 1.5 Good Product Evokes Humans' Kansei 7
- 1.6 Corresponding Development and Sales in Kansei 8

2. Kansei Engineering Case Study 11
- 2.1 Cars and Kansei Engineering 11
- 2.2 Brassieres and Kansei Engineering 13
- 2.3 Housing and Kansei Engineering 15
- 2.4 Kansei Engineering of Word Sound Image 19

3. Types of Kansei Engineering Technique 23
- 3.1 Kansei Engineering Type I 23
  - 3.1.1 Overview of Type I 23
    - 3.1.1.1 Step 1: Identification of Target 23
    - 3.1.1.2 Step 2: Determination of Product Concept 23
    - 3.1.1.3 Step 3: Breaking Down the Product Concept 24
    - 3.1.1.4 Step 4: Deployment to Physical Design Characteristics 26
    - 3.1.1.5 Step 5: Translation to Technical Specifications 27
  - 3.1.2 Application in Passenger Car Design 27
  - 3.1.3 Application in Brassiere Development 30
- 3.2 Kansei Engineering Type II 31
  - 3.2.1 Overview of Type II 31
  - 3.2.2 Application of Type II 34
  - 3.2.3 Consumer Decision-Making System and Designer Support System 36
- 3.3 Kansei Engineering Type III 37
  - 3.3.1 Hardware That Generates Sound 37
  - 3.3.2 Survey on Word's Semantic Differential 38
  - 3.3.3 Word Sound Image Diagnostic System 44

4. **Kansei Engineering Procedures: Kansei Engineering Type II** 49
4.1 Selecting Survey Target 49
4.2 Extracting Low-Level Kansei Words 52
4.3 Construction of Semantic Differential Scale (Part 1) 54
4.4 Primary Evaluation Experiment 55
4.5 Factor Analysis of Kansei Words 57
4.6 Extracting High-Level Kansei Words 59
4.7 Configuration of SD Scale (Part 2) 61
4.8 Secondary Evaluation Experiment 63
4.9 Preparation of Evaluation Samples 64
4.10 Extracting Item/Category 65
4.11 Statistical Analysis 68
4.11.1 Factor Analysis of Kansei Words 68
4.11.2 Analysis of SD Evaluation 68
4.11.3 Multivariate Analysis 69
4.11.4 How to Utilize Quantification Theory Type I 70
4.12 Creation of Databases 78
4.12.1 Kansei Words Database 78
4.12.2 Design Elements Database 78
4.12.3 Rule Base 79
4.13 Artificial Intelligence (AI) Prototyping 80
4.14 Establishing an AI System 80

5. **Kansei Engineering Type II Application Cases** 85
5.1 HULIS 85
5.2 FAIMS 88
5.3 Entrance Door Kansei Engineering System 89
5.4 Kansei Engineering System for Car Interior 92

6. **Hybrid Kansei Engineering** 97
6.1 Forward Kansei Engineering and Backward Kansei Engineering 97
6.1.1 Forward Kansei Engineering 97
6.1.1.1 Consumer Product Selection 97
6.1.1.2 Designer Product Development 97
6.1.2 Backward Kansei Engineering 98
6.2 Hybrid Kansei Engineering 99
6.3 Backward Kansei Engineering with Template 100
6.3.1 Image Recognition Procedure 101
6.3.2 Backward Kansei Engineering System 102
6.3.3 Actual Example of the Recognition Result 103
6.3.4 The Future of Hybrid Kansei Engineering 104

7. Virtual Kansei Engineering: Kansei Engineering Type IV 107
7.1 What Is Virtual Reality? 107
7.2 Virtual Kansei Engineering 108
7.3 Custom Kitchen Kansei Engineering 109
7.3.1 Computer Memory Content 109
7.3.2 Selecting Kitchen Style 109
7.3.3 Virtual Reality 110
7.4 The Evolution of Virtual Kansei Engineering 110
7.4.1 Fields Where Trial Products Are Either Expensive or Require a Long Design Time 111
7.4.2 Customer Decision-Making 111
7.4.3 Education and Training 112

8. Kansei Quality Management 115
8.1 What Is Kansei Quality Management? 115
8.2 What Are the Aims of QC and TQC? 116
8.3 Kansei Quality Management Thinking 118
8.4 Kansei Quality Management Case Studies 119
8.4.1 Kansei Quality Management for Restaurants 119
8.4.2 Quality Table for Hotels 121
8.4.3 Kansei at Ladies' Wear Department 123
8.4.4 Cash Register Work at Supermarket 125

Index 135

Preface

I began researching Kansei engineering (soft engineering) a long time ago. It has been gratifying to have companies the world over come all the way to Hiroshima to discuss actual implementations. In Japan, many actual examples have been developed, and many applications have emerged worldwide. In the academic field, I have been told that the Human Factors Research Group of the University of Nottingham has started research in Kansei engineering.

Originally, fuzzy things like sensitivity and emotion were ignored as objects of scientific study. There were only a few theories and little research, starting with the James–Lange theory (hypothesis on the origin and nature of emotions). Then the time came when the functions of the brain were technologically analyzed. It was in this era that *Kansei* was technologically reflected.

It all started when the idea to analyze the mind-structure of designers, which is somehow implicit, came to my mind. I wondered how designers express beauty in shapes. When I saw sketches drawn by designers, I thought they were extremely amazing. They were sensuous expressions, but I thought there must be some kind of rule behind them. These questions led me to work on what we call *item/category* analysis in this book.

Kansei engineering is similar to psychology in terms of grasping the image that exists in a person's (customer's) mind. It is related to humanity engineering in terms of translating the image into understandable design characteristics. It is an engineering discipline itself in terms of transforming the image into something that is measurable. Additionally, in regard to development strategy, that is, whether a product will be successful on the market, it requires involvement at the management level in handling the matter. Kansei engineering is thus a multidisciplinary science as well as a practical science.

Traditional product development has been performed based on the logic of business examples. Despite the saying that a product must be market-in oriented or customer-centric oriented, the product still has been designed from the standpoint of the company. Nowadays, though, costumers have become more mature, and their senses have become tremendously sophisticated. At this point, products that have been developed based on a company's strategy are far removed from the concept of Kansei. This is the reason customers have turned their backs on certain companies.

New product development begins with recognizing a product's potential role in improving the quality of consumers' lives and in supporting their lives. It is the role of the party who offers a new product to know the Kansei and lifestyle of consumers and to propose how to improve them. Kansei engineering is a new technology in product development that emerged to accomplish this role. Just as there are principles and rules in the beauty of design in the hands of designers, so also are there such principles and rules in the discipline of product design that match the consumers' Kansei.

From this point of view, Kansei engineering is recognized as "a mechanism that technologically translates consumers' Kansei into a product's design elements." It is a process in which the consumers' Kansei is first collected, and then its relation to the product design is determined. Then, a database or rule base is created that clearly defines the relationship, which enables product development to be performed anytime by referring to it. This has already been applied in the automotive industry, home electrical appliances, costumes, and construction machinery. It has even been undertaken in landscaping.

Kansei engineering is a technology that began at Hiroshima University. However, it is neighboring South Korea that has perhaps an equally strong interest in it. The South Korean government has ordered South Korean companies to introduce Kansei engineering technology by the early 21st century and to promote product development using this technology in an effort to outshine Japanese industry.

Indeed, Kansei engineering has great future possibilities, such as combining with virtual reality. My hope is that more and more people will become interested in Kansei engineering, read this book, and take part in Kansei engineering research and practical applications.

Mitsuo Nagamachi

About the Authors

Mitsuo Nagamachi, Ph.D., is the founder of Kansei engineering/Kansei ergonomics, an ergonomic new product development technology known and implemented worldwide. As a professor at Hiroshima University, Dr. Nagamachi created more than 40 new Kansei products, including cars, construction machinery, home appliances, brassieres, cosmetic products, handrails, toilets, and even a bridge over a river.

Dr. Nagamachi received his Ph.D. in mathematical psychology from Hiroshima University in 1963. He then studied medicine and engineering. From 1967 to 1968 he was a guest scientist at the Transportation Research Institute of the University of Michigan. Upon his return, he became the youngest ergonomic researcher appointed to Japan's Automotive Research Committee, whose mission was to make the Japanese automotive industry a world player. Dr. Nagamachi has consulted with the Japanese automotive industry on manufacturing, quality control, vehicle safety, management robotics, and Kaizen. In the 1970s, he began his research on Kansei engineering, which translates consumer's psychological feelings about a product into perceptual design elements. This technique resulted in the creation of numerous phenomenally successful products, including the MX-5 for Mazda, the Liquid Crystal Viewcam for Sharp, and the Good Up Bra for Wacoal.

Dr. Nagamachi has traveled extensively to teach Kansei engineering. He had served as a consultant in England, Spain, Sweden, Finland, Mexico, Taiwan, Korea, and Malaysia. In 2008 he was awarded the Japan Government Prize for the founding of Kansei engineering. He has received many academic awards from the Japan Society of Kansei Engineering. He has published 89 books and 200 articles.

Anitawati Mohd Lokman, Ph.D., is a senior lecturer in systems sciences at the Universiti Teknologi MARA, Shah Alam, Malaysia. She has worked for NEC C&C Japan and Toshiba Electronics manufacturing. She has considerable experience in computer systems and technologies, including Web systems. Her doctoral dissertation, which focused on the field of human computer interaction with the implementation of Kansei engineering, was directly supervised by Mitsuo Nagamachi.

1

The Emergence of Kansei Engineering

1.1 Developing a Product That Is Valuable for Consumers

Like the roll of waves, our economy booms and busts alternately. When times are good, consumers will be in the spending mood and buy products. It is during bad times that the value of a product is tested.

There is something in a good product that captures people's interest. A good product is more appealing to consumers in terms of its price as well as its function, shape, and color. It is a product that represents consumers' needs and has Kansei incorporated into it. Such products will sell even during bad times.

Some consumers' needs and Kansei do change with time, and some do not. Currently, consumers' Kansei is *products that have valuable content*, which means *good products that are comparatively inexpensive*. First of all, what does the word *valuable* or *good* mean? It means that the product is made from the consumers' point of view and to please the consumers. There are products that have been developed based on a company's perspective, which assumes that people will buy the product because they think it is convenient or reliable. Surprisingly, the company does not really understand the consumers' viewpoint, and this causes those products to fail in the market and appear to be disappointing. Developing products that get into the deep layer of what consumers actually want will enable good products and valuable products to be supplied into the market.

What does *undervalue* mean? It is not merely that the price is low. Most Japanese consider themselves middle-class and above. Additionally, due to the rapid increase in the standard of living during the last decade or so, and the optimistic mood of the bubble economy, the Japanese are imbued with a feeling of "classiness." Since this feeling does not diminish, they will not buy products that do not support their high standards of living. The Japanese have grown up with the sense that they live a life that makes a clear distinction between things that do not really need quality and things that require good functionality. Even though a product is good, if it has too many unnecessary functions, it won't capture the attention of today's consumers. Also, it does not mean that a product is good if it is inexpensive. The consumer's sentiment is, "I will buy the product if it is good and the price is reasonable for its quality." What is *good*, and what is *undervalue*? These are the two important issues for those who will be involved in future product development.

There are two directions in product development. One is the *product-out* concept, and the other is the *market-in* concept. The gist of product-out is that a company proactively produces and sells products they consider good, while for market-in, the idea is to develop products from the viewpoint of the market, that is, the consumers.

There were days when companies had substantially grown with the product-out concept, but those were the days when consumers did not have enough knowledge, and they were often naive when choosing products. Since then, many companies have attempted various ways to shift from the product-out concept to the market-in concept. However, they were not able to be objective in market-in product development, until today.

Nowadays, consumers' cabinets are flooded with goods. There is no more space to cram in new goods. Consumers themselves have become a great deal smarter. They have developed to a point where they consider such things as what makes them look beautiful, what improves their individual character, and how a product enhances the value of life. If future product development does not strategize the market-in concept, consumers will turn their backs on it.

We incorporate Kansei to win over consumers. We used to hear the term *user friendly*. In recent years, the terms *consumer-centered* or *human-centered* have emerged. In the years to come, Japan will face a tremendous challenge due to its aging society. It's a known statistic that, in the near future, one out of four persons will be over 65 years old. Development of products that are easy to use and appreciated by people, including the elderly, is human-centered product development. *Human friendly* is a term that describes this. In the future, environmental or similar problems will also be related to product development. The trend of product development that carefully considers the well-being of humankind and harmoniousness with the entire world and the environment will become focal.

1.2 The Emergence of Kansei Engineering

Kansei engineering is a technology that unites Kansei (feelings and emotions) with the engineering discipline. It is a field in which the development of products that bring happiness and satisfaction to humans is performed technologically, by analyzing human emotions and incorporating them into product design.

Sometime around 1970, I visited manufacturing companies regularly. I could see that people were being encouraged more and more to make purchases due to the healthy economy, and companies responded with mass production of goods. However, I foresaw that people would stop buying when their cabinets became overfilled. If this happened, perhaps only the good product, that is, the good-quality product, would win over consumers. However, since the good-quality products are long lasting, the market flow would then become slow. Then, it crossed my mind that perhaps only the products that consider people's feelings, that is, the products that people really want and desire, will sell. Then, there will be an era when products are made to express one's personality. Thus began the Kansei era, in 1970, followed by Kansei engineering research.

The flow of the emergence of Kansei engineering is shown in Figure 1.1. The left column of the figure shows the consumer's desire, while the right column shows the corresponding activity at a company. Companies must develop products that correspond to the change in people's feelings and emotions in order to survive. They should employ a product development strategy that anticipates people's feelings and emotions. The Kansei era emerged about 20 years after I first started to ponder it, and it is predicted to continue indefinitely. Maybe it will continue forever, or perhaps it is more reasonable to think that the form of Kansei expression will change in accordance with the changing times.

Let's say we name the product that was produced using Kansei engineering technology the Kansei product. The Kansei product is not an expensive product or a high-end product. It is also not something that emphasizes good looks, appearance, or style. The Kansei product refers to that which can actualize the needs and emotions, considering functions and shapes, and

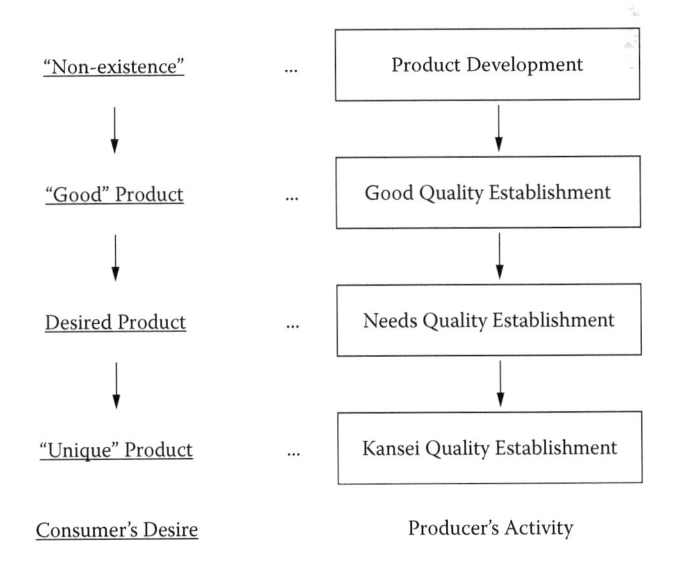

FIGURE 1.1
The concepts underlying the emergence of Kansei engineering.

FIGURE 1.2
The purpose of Kansei engineering.

also whether the consumer would want that product and what the product offers.

For example, we can realize the product that elicits Kansei simplicity in a video tape recorder (VTR) that has buttons that are clearly visible and makes recording easy, or a telephone that does not have complicated functions but enables the user to perform necessary functions easily (Figure 1.2).

In implementing Kansei engineering, it is important to investigate the people's Kansei. When Kansei information is properly collected and analyzed, it can then be translated into a technical design. Even in the engineering field, there are cases where new technology will be required in order to actualize Kansei. In such cases, many patents and models of new techniques will appear. Product development that utilizes Kansei engineering will not only produce diverse products that are friendly to people but also be an impetus for new technology development. In this sense, we could also say that Kansei engineering is a new technology in the new era.

1.3 What Is Kansei?

Before we consider the main subject of Kansei engineering, let me first explain the principles of Kansei.

According to the *Shin meikai* Japanese dictionary by Kindaichi Kyosuke et al., Kansei is "intuitive mental action of the person who feels some sort of impression from an external stimulus." In the psychological definition, Kansei refers to the state of mind where knowledge, emotion, and passion are harmonized; "people with rich Kansei" are full of emotion and passion, and able to react adaptively and sensitively to anything.

The term *Kansei* used in Kansei engineering refers to an organized state of mind in which emotions and images are held in the mind toward physical objects such as products or the environment. For example, concepts such as luxurious, elegant, flashy, and young, as in "that product is a bit luxurious and elegant," or "those clothes are flashy and young looking," are all Kansei impressions of products. The term Kansei used in Kansei engineering in most cases takes the form of an adjective, but it may also be a noun, as well as foreign words written in *katakana*.

However, since Kansei is all about the image held in one's mind, it is therefore expressed using various media, such as words, facial expressions, or drawings in order to make others understand. It is still unknown whether the expression itself matches the image in the other person's mind. This is an issue to be considered. From the beginning, the difficulties in measuring something like Kansei have been apparent, since it deals with a person's mental attributes. We ask people to express an image using common, everyday words. There is no English word that translates exactly the meaning of Kansei. If it still has to be translated, *psychological feeling* can be used. However, since this term causes more confusion, today the original Japanese word—Kansei—is used. Likewise, we use the term *Kansei engineering* in English. Sometimes in Kansei engineering, people are asked to express their Kansei in words upon seeing products, or regarding products they want to buy in the future or products that are not yet available. These are called Kansei words.

In Kansei engineering, some Kansei terms reflect the times and do change occasionally, such as trend-related Kansei, while others virtually do not change at all, such as fundamental Kansei (colors, etc.). Additionally, cultural differences among countries cause differences in the Kansei itself, and some Kansei are similar and yet still different in the expressed Kansei words. Careful attention is required when applying Kansei engineering in this matter.

1.4 Kansei Is Something Comprehensive

When you say, "It is an elegant dress," upon seeing someone's garment, you feel a Kansei of *elegant* as your impression of the whole dress. However, when we talk about fashion design, the impressions for one-piece and a suit will be different. Furthermore, type of collar, number of buttons, and pocket design will give different impressions. For skirts, depending on what types they are and even the material used, the overall implied Kansei will be different.

When we think of an article of clothing, we can imagine the breakdown of its elements (or parts) such as (1) overall style, (2) upper piece type, (3) collar style, (4) numbers and position of buttons, (5) pocket design, (6) type and length of skirt, and so forth. The composition of these elements will elicit certain Kansei. A slight difference in buttons will generate different Kansei. Different numbers of pleats in a skirt will elicit different Kansei. Much Kansei exists even within the elements themselves, and each individual element affects the overall Kansei.

Besides being an overall or comprehensive impression, the elements of Kansei are the subject of Kansei as well. Therefore, in performing product development using Kansei engineering technology, the following procedure is necessary:

1. Break down the design into separate elements.
2. Interpret the Kansei of each element.
3. Design the overall product.

It is important to grasp in advance which of the dozens of elements are having a great influence on the overall Kansei, and then to pay attention to those elements and incorporate them into the product design. This is illustrated in Figure 1.3.

Let's say, when we break down the elements of product A, we identified elements a_1 through a_7; and from a statistical analysis of those elements, we found that a_5 and a_7 greatly affect certain Kansei. This is called degree of contribution. This means, in order to incorporate the specific Kansei in product A, we should incorporate the elements a_5 and a_7 into its design.

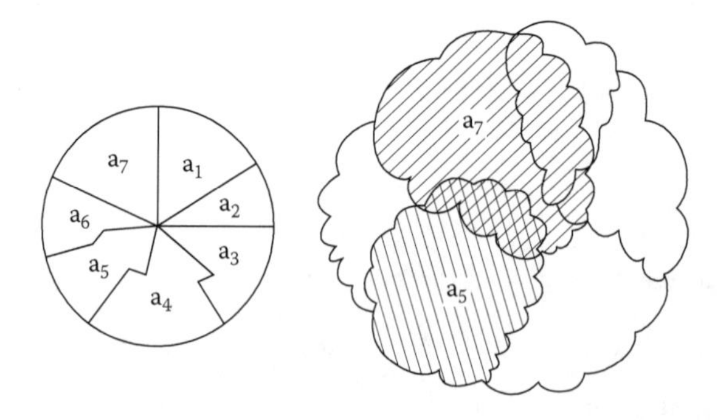

FIGURE 1.3
Kansei is something like the right figure, the element as a whole with indistinct boundaries, rather than the left figure, a mosaic of elements.

1.5 Good Product Evokes Humans' Kansei

Good product refers to a product that is designed to matched the hidden Kansei signature of the consumers. When consumers find such a product, they will be very impressed and feel, "Wow, what a great product!"

If 60–70% of the Kansei product matches with the consumers' Kansei, and the balance of 30–40% of the Kansei is in the realm of something new and excellent, the consumer will become very impressed and believe that it is an excellent product. Of course, this will be a motive for buying. The 60–70% of Kansei can be analyzed in a Kansei engineering study, and the remaining 30–40% will depend on the creativity of designers and research and development (R&D) personnel. This kind of percentage is important. For example, even if the product matches 80–90% of the consumers' Kansei but could not elicit a feeling of novelty or creativity in the consumer, it will be perceived as a clichéd product. Conversely, if the portion that matches the consumers' Kansei is 10–20%, and the balance of 80–90% is the result of creativity from R&D, it will be perceived as a future-oriented product, which consumers do not want to have around. Concept cars or costumes that appear quirky belong to this category of products.

Product R&D personnel must keep their sights on the product development that evokes consumers' emotions. In order to do that, they must strategize to concentrate on grasping the content of the consumers' Kansei at the current moment.

However, they should not forget that a product that evokes consumers' emotions also lifts consumers' Kansei to the next level. Consumers will develop a kind of Kansei intelligence as they come into contact with the product, and the Kansei level will develop. Therefore, the next level of product development must tune in to the increased Kansei level. Otherwise, the product will fail to evoke consumers' increased Kansei, and the new product will disappear from the market. In this sense, product development is ultimately a kind of battle to improve consumers' Kansei level.

Good product development means developing a product that evokes consumers' Kansei. Also, this makes consumers somehow attuned to the Kansei and causes improvement in their level of Kansei. We need to be careful not to forget this, because forgetting may cause failure in the strategy to continuously evoke consumers' Kansei in the next product development after a market success, thus causing the product to disappear completely from the market.

There is another general rule related to evoking consumers' Kansei. In the discipline of psychology we have the Weber-Fechner law. As shown in Figure 1.4, making the vertical axis the Kansei increment and the horizontal axis the stimulus increment, in order to increase the same value of Kansei increment, ΔK, while the stimulus is small, increasing it at almost the same portion will result in an increase of ΔK. However, when the Kansei level becomes higher and stimulus is not increased, we will not get the ΔK

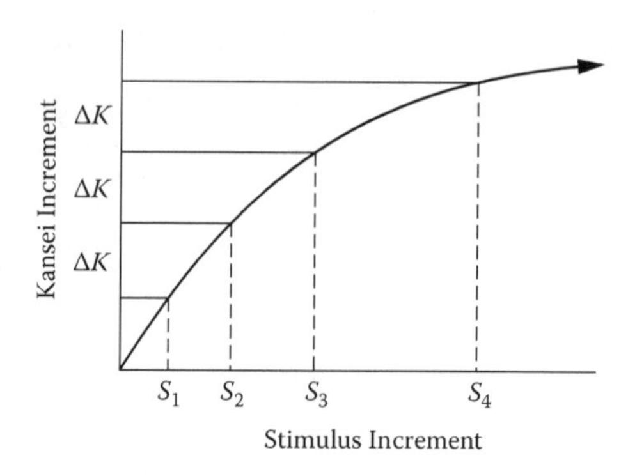

FIGURE 1.4
Relation between stimulus (device) and Kansei increment.

of Kansei increment. In other words, between Kansei (K) and stimulus (S), there is a logarithmic relation, as described in the following:

$$K = C \log S \text{ (C is a constant)}$$

This is the Weber-Fechner law.

In the field of Kansei engineering, this can be described as follows. Let's say we have product B. It can be a television or a refrigerator or whatever you wish to imagine. I want to raise the Kansei level of product B one rank higher, which is ΔK, to impress consumers and thus contribute to sales. So, let's assume that I have increased the stimulus from S_2 to S_3 and achieved the target. This means that I have added a few functions or made it look nicer. Of course, since the development costs have increased as well, the price will be higher.

After a few years, if we want to increase ΔK of Kansei for the same product and evoke the impression again, we have to make an even greater effort than last time to increase the stimulus from S_3 to S_4, and throw extra investment into development. If we want to differentiate the Kansei point of high-end television, such as Gao or Teio, from conventional televisions, we have to produce *big* Gao or *big* Teio, otherwise consumers will not be impressed.

I hope you understand how complicated it is. After some time, a phenomenon occurs: consumers turn away from the product. It is not an easy task to develop Kansei products that impress consumers all the way since they will soon learn and become smarter. We should not forget the Weber-Fechner law. It also works in the product's Kansei.

1.6 Corresponding Development and Sales in Kansei

If we look at previous bestsellers, we will have a better understanding of the consumers' Kansei, and incorporating it into a product will make it sell well. For example, Sharp's camera-type VTR LCD ViewCam is a product that made the company's VTR share jump from only a few percentage points to 20%. This was the result of extensive analysis of consumers' four dimensions of pleasures: shooting pleasure, watching pleasure, face-to-face pleasure, and the pleasure of playing on the spot. The results of the analysis were then incorporated into the product's design technology.

What I want to emphasize here is that if we only focus on matching consumers' Kansei in the development of the product itself without correspondingly considering the point of sales, we will fail, even though we produce a good product. For example, when the homemade bread machine first launched, all the newspaper promotions that it would surely become a hot seller tremendously helped sales. This is because the product had incorporated the consumers' Kansei of *gourmet consciousness* and *premium feeling*, and consumers could then bake delicious breads at home. However, after a while, the sales suddenly stopped. Upon investigation of the root cause, companies found that, since the flour was sold at electrical appliance stores and was packed into unattractive boxes, this contradicted the product's image of stylish housewives, making them resistant to buying. Consumers' Kansei had been effectively utilized in the product development but not to the distribution point. On the other hand, the product was selling well in the United States.

Asahi Beer has achieved great success with its Super Dry line because Kansei has been utilized correspondingly in both product development and sales. R&D personnel focused on developing a new beer. They sensed that the Kansei of consumers' taste were *robust* and *crisp* and confirmed this with a survey of 5000 consumers. Since the two tastes contradict each other, it was very difficult to actualize the product. However, the brewers perfected a skill that smartly uses yeasts related to each taste to accomplish the Super Dry.

Before that, Asahi Beer's market share was as low as 8.9%—at the rock-bottom level—and it was promoting a company-wide corporate identity (CI) movement for a revival. To achieve that, they had launched a new product campaign all over Japan, starting simultaneously from Sapporo and Kagoshima. At that time, their general employees banded together and focused their efforts on sales. A CI committee member suggested that if they buy all of the old Asahi beers from vending machines around the city, the new product will reach consumers faster. So, the employees hunted all the old Asahi beers on Saturdays and Sundays. As a result, the general consumers got a taste of Super Dry earlier than they otherwise would have, and there they outshone all other companies. Perhaps the only drawback was that the employees also consumed the old beers they had bought!

It is necessary to take corresponding actions in Kansei implementation between the product development point and the sales (marketing) point. Automakers were the first to implement Kansei engineering intensively. Even though they have come out with excellent products based on

Kansei, it is uncertain whether the car dealers are practicing sales activities correspondingly. When customers are looking at a new product and they show interest in a certain part, if the salesperson can explain how the R&D personnel made an effort to include Kansei in that aspect, it will touch the hearts of consumers (Figure 1.5).

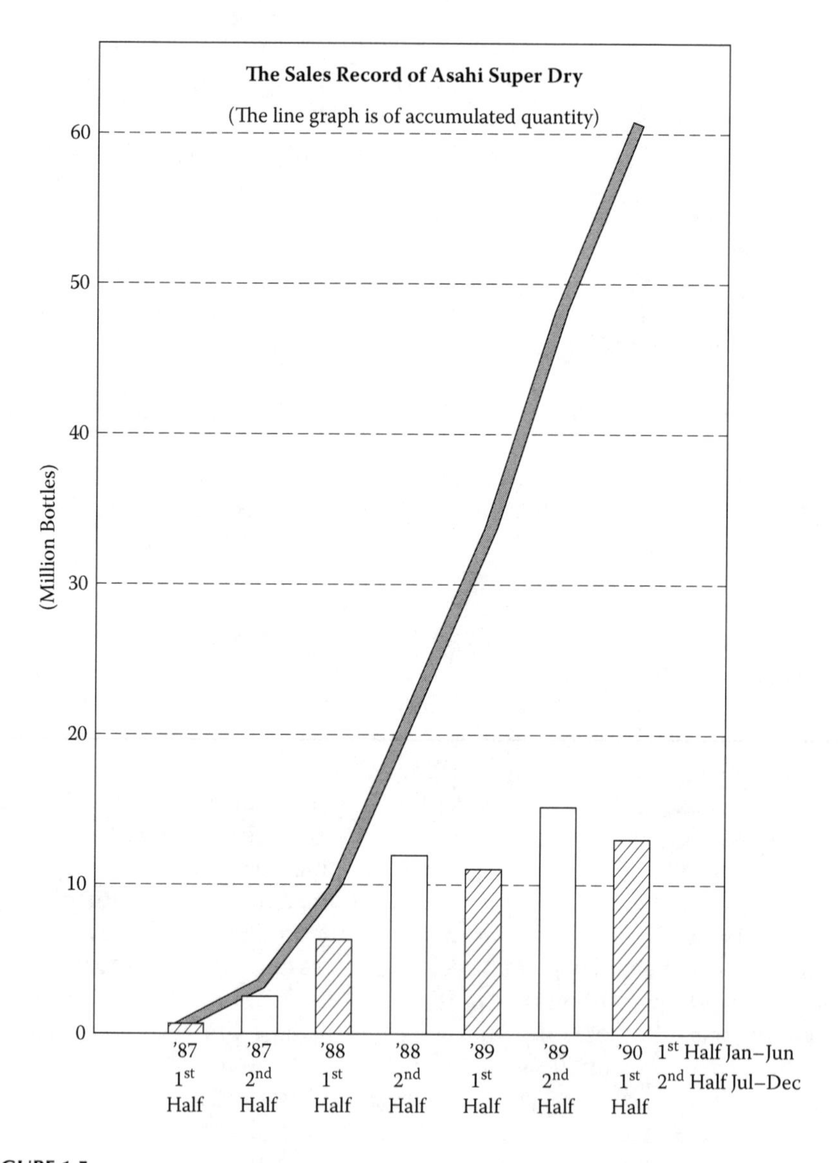

FIGURE 1.5
Volume of sales extended because sales activity is corresponding with Kansei product.

2

Kansei Engineering Case Study

2.1 Cars and Kansei Engineering

Let's discuss some examples of Kansei engineering in products around us and the points at which Kansei engineering was applied to these products. Let's start with cars. It is said that Japan's economy has been powered by cars and home appliances. Supporting industries are especially broad and diverse for cars—ranging from steel as the raw material to electronic parts. This represents a big contribution to Japan's economy if the products are selling well. It is a source of great pride for me that Kansei engineering contributes to this success.

Kansei engineering was implemented in car production starting with Cima on through Presia for the Nissan Motor Company, which led the innovation with a totally new car design. For Mazda, it started with the development of Persona Interior, which was also called a *moving sofa*. This is a product that was born from a development concept under the motto of "Interior," which shows the great value of Kansei engineering in interior design. One very good example is the Eunos Roadster (MX5), a product that was developed later using Kansei engineering in every inch of its design. Mitsubishi Motors was the first car maker to implement Kansei engineering, especially in the research of its vehicle compartment. Diamante is a product of such research. Toyota and Honda also have studied Kansei and have applied it to their designs.

Outside Japan, Italy's Fiat and America's Ford have great interest in Kansei engineering research. Ford utilized it in its Taurus model. Fiat also has begun to make changes in the style of the latest passenger cars. Porsche, its popular German car maker, has also started research on Kansei engineering. In Korea, Hyundai Motors and Samsung Electronics are already quite advanced in Kansei research.

I'll explain the relation between Kansei and car design by example.

First, the exterior (appearance, style) design. When we studied the Kansei of luxury passenger cars, we found that the height of the front hood plays a very important role. In other words, if customers want a luxury car, then we need to do something to make the front body look a little bit higher. However, it is also a fact that if we make the front grille higher, air resistance will be greater. Therefore, making the front body design appear higher and yet be streamlined is important. If you look at the design of a Mercedes front hood, I think you'll be able to understand the relation between Kansei and design. The same goes for Japanese cars, where we can see that luxury cars tend to have a higher front body. In addition, it is quite obvious that cars designed to appeal to young people have a lower front body, while cars for older people have a higher one.

Next, regarding Kansei with respect to interior design, let me explain the term *spacious feeling*. Passenger cars typically have a rather small passenger space. Therefore, it is necessary to make this small space look spacious. Our studies showed that there is a specific rule in Kansei design for making a small space look spacious. Figure 2.1 shows a part of the output from the IKDES (Interior Kansei Design System) computer system, which displays the result of the rule. I'll explain some of the main points: (1) The distance from the driver's eye to the tip of the dashboard should be made a little bit longer; (2) the distance from the eye to the base of the windshield should be made longer; (3) the horizontal space for the instrument panel in front of the driver should be wide; (4) looking from the eye point, the distance between the windshield base and the front of the roof should be long; (5) the width of the center console (control panel for radio, etc.) should be wide.

Besides these, there are a few combinations of design elements to make the driver's seat area look wide. Auto makers can incorporate the combination to make their car interior look spacious and match the concept of the car. On the other hand, for a sports car design, incorporating a design that induces *narrowness feeling* will make it feel more like a sports car.

If a car interior is made larger physically, the spacious feeling will become more apparent, but for conditions that are subjected to physical constraints, as passenger cars are, it is possible to create a spacious impression psychologically and emotionally by putting some effort into interior design. Kansei

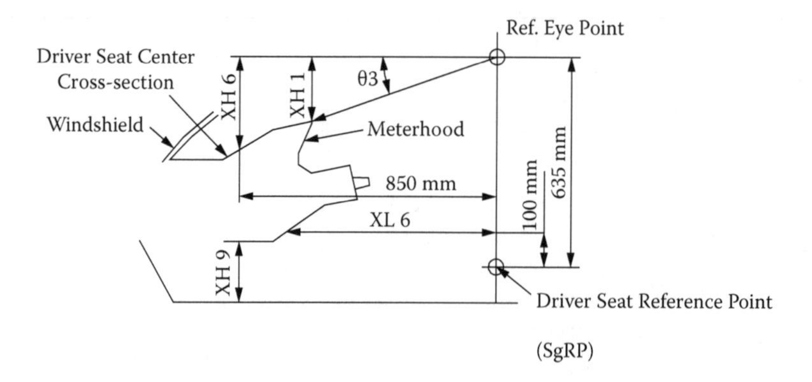

FIGURE 2.1
Part of the output from IKDES—the system used in diagnosing a car interior's *spacious feeling*.

engineering is useful in determining such design conditions. Similarly, Kansei engineering is being used in various Kansei applications to make car interiors look luxurious and premium.

As described earlier, since cars are expensive and consumers need to drive them for pleasure and various other purposes, it is important to analyze consumers' Kansei and incorporate it into every part of car design.

2.2 Brassieres and Kansei Engineering

Wacoal, a top Japanese manufacturer of women's intimate apparel, has produced many bestselling products. Especially in its brassiere line, Wacoal has produced one bestseller after another, such as the Hanakin brassiere, which "charms young women's feelings," and Kokochi E, which embeds a memory alloy.

The product I'm going to describe next was a phenomenal success, and it was developed out of a fresh dimension of consumers' Kansei. According to a survey of 2000 women who participated in a product trial to determine what they expect when wearing a brassiere, the answer was they *want to feel beautiful and elegant*. Upon further questioning about what conditions are required to achieve the expectation, there were various opinions. From the tally sheet, the top two opinions are as follows:

1. Both breasts should rest within the chest width. Breasts that exceed the edge of a woman's narrow chest will surely result in a plump look, and this contradicts the image of beautiful and elegant.
2. The nipples should project out from breasts nicely and in parallel. Furthermore, they should face a little bit upward evenly. If the two nipples face left and right separately, they destroy the elegant feel. Facing a little bit upward will portray an image of youthful beauty.

A brassiere that satisfies these two conditions will be a product that is purchased by many women who desire to have beautiful breasts. Since they have had experience in the past with the Three Quarter Cup, a phenomenally successful product that was labeled as a slightly sexy brassiere, this time the maker cut the brassiere to a V-shape to make breasts look bigger and wider (Figure 2.2).

This is how they determined the design conditions to realize the Kansei of women wishing for beautiful breasts. It should be a simple matter to proceed to the next steps, which is selection of material, and to specify design elements to complete the product. However, women's nipples are sensitive body parts. They are as sensitive as fingertips, if not more. If they are squeezed

FIGURE 2.2
The brassiere Kansei engineering.

even a little, the person will experience a tight-fitting sensation and will not use the brassiere. To satisfy the abovementioned design conditions:

1. in order to make breasts settle within the body chest width, the two breasts need to be drawn toward the center.
2. to make them face upward evenly, the bra must be designed to a shape that uplifts the whole breast.

In addition, the product must not give a tight-fitting feeling and yet must be comfortable against the skin. It sounds impossible to realize such contradicting conditions. Wacoal's R&D personnel reached the following conclusions. To make the two nipples face upward evenly, the breasts should be drawn toward the center without causing any discomfort, and they must be pushed up from below. The design must consider gentleness to women's sensitive skin. Then the R&D personnel performed research on fibers that are gentle to the skin and yet strong. They were already familiar with a gentle and strong amorphous fiber used in an earlier successful product, so they performed experiment after experiment to make the fiber finer and turn it into fine textile. Finally, they succeeded in creating a fiber that suited the purposes. This is the product development story of the Good Up Bra. This product got its name quite simply from the act of drawing two breasts together and pushing them up.

An article in the *Nikkei* on December 21, 1993, highly praised the bra's sales performance, which had expanded from 3 million to 7 million sales of that single item by 1995. This amazing story tells us that product development with its perspective tuned to both the Kansei and the needs of consumers will surely succeed.

2.3 Housing and Kansei Engineering

In 1970, Kansei engineering (which at that time was called emotion engineering) research was initiated by the engineering faculty of Hiroshima University—the first instance of this kind in the world. It was inspired by a consumer's grievance claiming the house that was built for him was different from what he had initially imagined. He had worked very hard for many years to save money and when his savings reached the target amount, he cheerfully went to a construction firm and conveyed what he had in mind to the architect. Based on the information, the architect came up with a design draft corresponding to the architectural standard. The client noticed that the draft was quite different from the initial design he had envisioned, but he calmed his fears by saying that he was dealing with an expert in the trade. He was shown wall and bathroom tile samples so small that he had to choose mostly from his imagination, but things turned out to be totally different from what he had imagined. Almost everything turned out to be different from his initial image, and while living in the house, he became infuriated with the result.

While listening to his grievances, I noted that among the underlying causes were the following:

1. The image in the consumer's mind was not conveyed to the architect in the correct language.
2. The architect took his pride as an expert in house designing too far, thus trying to fit the needs (images) expressed by the consumer into his implicit expert design template.
3. The consumer gave in from what he had initially insisted on due to his feelings of inferiority as an amateur.
4. The samples used to assist the consumer's images were too small, resulting in a difference in the actual completed project.

In other words, there was no technique involved in conveying the desired images in the consumer's mind to the architect, and there was a lack of tools to describe the consumer's image objectively. That motivated me to conduct research on engineering techniques that can objectify consumers' Kansei into designs.

First, we prepared an approximate miniature house, where the colors of its ceilings, walls, and floors can be changed freely. We asked subjects to look at it and to evaluate their response using SD (semantic differential) Scale (see Chapter 4). At that time, we compared the evaluations of subjects from final-year and postgraduate students of the Faculty of Architecture with other general students. We found that the architect-to-be subjects had properly evaluated the Kansei, and based on the data, we performed the Factor Analysis and Quantification Theory Type I. Using the result, we created combinations of Kansei for interior decorating and materials such as wallpaper that could be used in the actual interiors. This was a great success. There was a rapid increase in customers within a short period of time. In addition, while a salesperson generally requires a few days to discuss plans with a consumer and then produces several drafts before the consumer is satisfied with the plan, using the Kansei engineering system, the consumer's opinions are addressed on the spot and can be confirmed with a 1 m × 1.5 m sample. In many cases the deal was closed within the first two hours. This greatly contributed to profit-making by dramatically reducing marketing costs. Since there is a psychological phenomenon where the brightness of color will increase if the palm-wide sample is magnified to a wall-sized one, this new idea of "the bigger the sample, the better" also contributed to a favorable situation, and there were no complaints.

These days, the functionality of computers is dramatically increasing, while their price is inversely going lower. This has made me think that if only we had utilized computers for Kansei engineering 10 years ago, we could have provided even greater efficiency than the manual system. It was the beginning of the present Kansei engineering system. The procedure followed 10 years ago was:

1. Collecting Kansei words. We collected about 600 words by visiting home shows and recording the conversations between salespersons and visitors, and browsing through magazines and selecting words related to interior design and appearance. We reviewed those words, omitted any overlapping ones, and developed the SD Scale by making pairs of antonyms such as *elegant—inelegant*.
2. Slides of houses. We visited home shows or general houses, and while categorizing a few first-class architectural styles and types of designs (described later as item/category), we selected houses in all categories and took pictures of them to make slides.
3. Evaluation experiment I. Using the 600 words that had been transformed into the SD Scale and 20 sets of randomly picked slides, we performed evaluation experiments with students as subjects. The purpose was to reduce the number of words to fewer than 100.

We performed factor analysis of the evaluation result and chose a few words from each factor. Finally, we compiled a list of 40 words.

4. Evaluation experiment II. Using 40 slides, we evaluated the house images with 40 Kansei words using the SD Scale. In this experiment, we made the evaluation possible for the design of seven important elements; appearance, structure, entrance, Japanese-style room, Western-style room, kitchen, and bathroom. The subjects were 40 architecture designers. From the result of this evaluation experiment, we completed a database for the seven elements.
5. Computer system. I thought of creating a computer system that has a judgment function similar to that of humans by using these databases. The judgment function is called Expert System.

The image that I had in my mind is like this. Let's say a consumer who wants to build a house visits a construction firm. He has an image in mind such as, "I want to build a house like this…." First, regarding its external appearance, let's say he has imagined it to be a big and elegant Japanese-style house. So, we input *big, elegant,* and *Japanese-style* into the computer. The computer can understand what these words mean, and it also knows what contributes to a design that is *big*. It will add the meaning of the other two words, build a visual of the appearance, and display it. If the consumer is satisfied with that, we will then proceed to the entrance, and in the same way, we will ask about the consumer's image, and then input those words. We will continue in sequence (Japanese-style room, Western-style room, kitchen, bathroom, and structure) to input the images that he has in his head and show the image created through the judgment (known as *inference*) by the computer. If the consumer is satisfied with all the outputs, the result will be expressed in the form of an architectural plan.

In order to create this image, I had prepared (1) a database that can understand the meanings of Kansei words input into the system, (2) an inference function with rules about what kind of design relates to each Kansei word in each design of the seven elements of a house, (3) a design database that will turn the inference result into an image with shapes and colors, and (4) a control system that manages and operates the whole system by following a set of predefined rules.

I named this system HULIS (read as *hyu-lis*). Examples of its output are shown in Figures 2.3 and 2.4. HULIS was approved by the Ministry of Construction as a tool for interior coordinators, and it received the Year 1985 House Building Premier Award.

HULIS was the first application of Kansei engineering that was turned over to a computer system in the form of the Expert System. As will be explained in a later chapter, one Kansei engineering application after another that takes advantage of the computer's ability has been established.

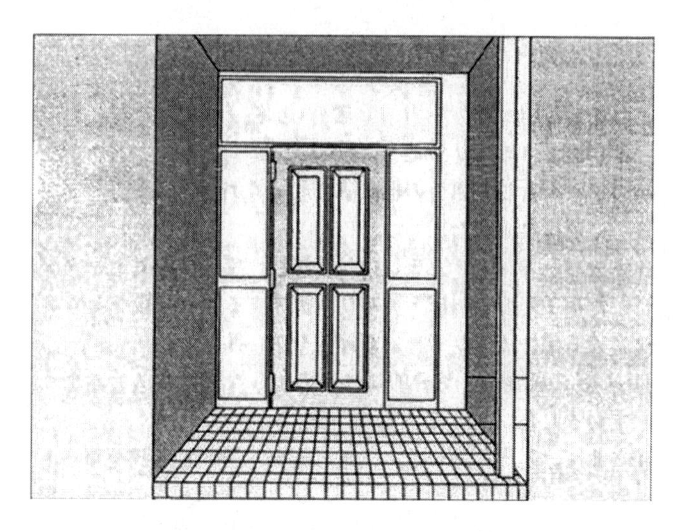

FIGURE 2.3
Part of HULIS (entrance).

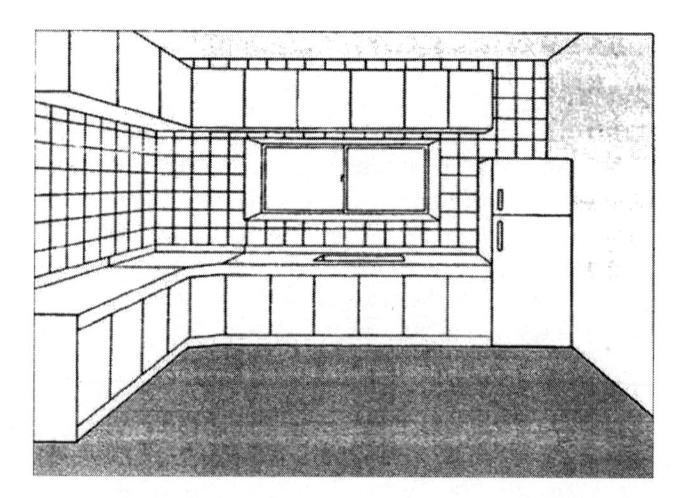

FIGURE 2.4
Part of HULIS (kitchen).

For HULIS development, the basic image of the completed system that I envisioned is as follows:

1. First, the person who wants to build a new house will bring the land map to a construction firm. The computer will then trace the land map and calculate the area. At the same time, it will perform the inverted shadow calculation in three-dimensional space to determine the effective area the house can be built on.

2. The client will describe the lifestyle he envisions in the new house, the number of family members and their ages, and the budget. All this information will be input into the computer.
3. The computer will automatically process this information and design the house in single or double story and room arrangements.
4. Using the house design that the computer has generated, Kansei will be input for the interior such as appearance, structure, entrance, Japanese-style room, Western-style room, kitchen, bathroom, and others similar to HULIS. Then, the computer will shape the image of each room and area accordingly.
5. When all these have been decided, the computer will automatically display the floor plan and path from three sides in color. It will then generate the three-dimensional graphic for each room. At the same time, it will also generate a quotation.
6. Then, the client will show the color diagrams to his family, and they will discuss the features. If they are satisfied, they will place an order.

This is called the Auto-Design HULIS System, and it is halfway completed as of the writing of this book. Since this system is equipped with designing capability and a Kansei-understanding system similar to an architect, this mechanism will generate the most suitable design for the client's lifestyle as well as a refined design based on the designer's Kansei. Thus, a design that satisfies both the client and the architect can be developed. This is a perfect creative activity in Kansei work.

2.4 Kansei Engineering of Word Sound Image

Sounds such as *ooh* when we are surprised or *hu hu hu* when we laugh are not meaningless or random. The theoretical evidence for this will be explained in detail later. Our throat structure (hardware) and how we move our mouth and teeth, as well as how we breathe, affects the formation of many kinds of words that come from our mouths. Each of these words is properly determined by the hardware that produces the word and its operation (software). This is what we call *phonology*. Additionally, it is a fact that whether we write in *Kanji* (Chinese character) or in *hiragana* (Japanese character), the sound of each word has its own Kansei.

For example, the sound of the word *ki* (き): If we just simply imagine the word, it gives a sense of insensitive and stiff, while for the image of the word *ru* (る), we can feel a sense of roundness and smoothness. As just described, each word has its own kind of feeling. This is what we call Kansei engineering of Word Sound Image. When a baby is born, we ponder for a few days before giving a name to the baby. We take the time to decide which name to give, since we are obsessed with the number of strokes and the meaning of the Chinese characters to be used. Actually, the Kansei of a name lies in the feel when it is pronounced, which is cheerful or energetic.

If we can evaluate the monophonic letters like *a*, *i*, *u*, and so on, and apply some quite complicated mathematical processing, we may be able to create a system that can diagnose the Kansei we feel for any kind of word. A computer system called WIDIAS (Word Image Diagnosis Fuzzy Expert System) was developed for this. Table 2.1 shows the result when the image for the word *ma-chi* (マーチ) was diagnosed using WIDIAS. In the table, the first letter is *ma* (マ), the second letter is "-" (ー), and the third letter is *chi* (チ). In the left column, antonyms of adjectives like *soft–hard* are listed. If the numerical value is greater than 0.5, it shows the left side of the adjective's pair (in this case, soft), and if the numerical value is smaller than 0.5, it shows the right side of the adjective's pair (in this case, hard).

As for the word *ma-chi* (マーチ), the elicited Kansei is shown in Table 2.2.

TABLE 2.1

Kansei Evaluation for the Word *Ma-chi* (マーチ) Using WIDIAS

Input Letter String [Ma-chi] (マーチ)

Adjective Pairs		First Letter	Second Letter	Third Letter	Overall Judgment
Soft	Hard	0.7019	0.5385	0.4338	0.53846
Bright	Dark	0.6635	0.8461	0.5481	0.64380
Broad	Narrow	0.5096	0.6635	0.4135	0.41240
Unique	General	0.5289	0.5000	0.5577	0.50000
Expansive	Unexpansive	0.5192	0.7308	0.4327	0.62580
Heavy	Light	0.4172	0.2788	0.3077	0.36780
Refreshing	Old	0.4808	0.7404	0.4423	0.59170
Clear	Unclear	0.5192	0.8750	0.6827	0.63590
Simple	Complicated	0.5000	0.8365	0.5577	0.83654
Glamorous	Unglamorous	0.4904	0.6250	0.4712	0.48100
Warm	Cold	0.6923	0.6731	0.4519	0.67308
Individual	Common	0.6154	0.4423	0.5865	0.51430
Have uplifting feeling	No uplifting feeling	0.4423	0.7115	0.5192	0.63560
Nice ring	Ill-sounding	0.5769	0.8461	0.5000	0.45380
Roundish	Squarish	0.7019	0.5289	0.3173	0.52885
Gentle	Unkind	0.7308	0.6539	0.4327	0.50060
Masculine	Feminine	0.3558	0.4808	0.5096	0.48077
Have sense of flowing	No sense of flowing	0.4808	0.5769	0.3750	0.72490
Sharp	Dull	0.3365	0.6250	0.6923	0.57120
Powerful	Powerless	0.4808	0.6539	0.4423	0.65385

TABLE 2.2

Kansei Characteristics for *ma-chi*

ma (マ)	"-" (ー)	chi (チ)
① Gentle	① Bright	① Sharp
② Soft	② Nice sound	② Clear
③ Rounded	③ Simple	

The overall judgment for the three letters taken together will (1) be simple, (2) have a sense of flowing, (3) be bright, (4) be warm, and (5) have an uplifting feeling. Nissan Motor's compact car was named *March* (マーチ), and the style of the car attunes nicely with the image of the word *ma-chi* (マーチ). Most probably, the name was given by Nissan Motor's R&D personnel after they had considered many possibilities. However, the WIDIAS database can perform such a diagnosis just by combining the sounds of each single letter in the word, totally unrelated to the car produced by Nissan.

When we try to use the Kansei engineering technique, we know that there is a specific principle even in the image of a word that we use unconsciously, and we actually have images of word sound. This will be explained in detail later.

3

Types of Kansei Engineering Technique

Kansei engineering research began at Hiroshima University around 1970, but in recent years much more research has been performed. At the moment, Kansei engineering methods have been classified into three types as described below, but new methods will be introduced in the future.

3.1 Kansei Engineering Type I

3.1.1 Overview of Type I

Kansei engineering Type I is the easiest to understand and introduce. The well-known KJ Method can also be used. This method breaks down a targeted product concept into a more detailed concept, and while expanding it to several levels, it will be interpreted in terms of the physical characteristics of the product design. Figure 3.1 illustrates this concept. Steps in carrying out this method can be described as follows.

3.1.1.1 Step 1: Identification of Target

Identifying the target market group includes determining to whom the product is meant to be sold and how to handle those people's Kansei. First of all, decide on the market target group: for example, children or adults, or young women or young men. This can be determined by the experts in the company's product field, or from the results of a marketing survey. Make a decision based on sufficient survey data and ample studies.

3.1.1.2 Step 2: Determination of Product Concept

When the target market has been decided, we then have to decide what kind of product concept should be incorporated into the product that is to be developed for the customers. Of course, this should be done by surveying and studying the target's lifestyle and other aspects. You can determine the concept by consulting skilled and experienced R&D personnel and by collecting ample data. For example, if the target group is young girls, collect data about their lifestyle and trends and how these relate to fashion, as

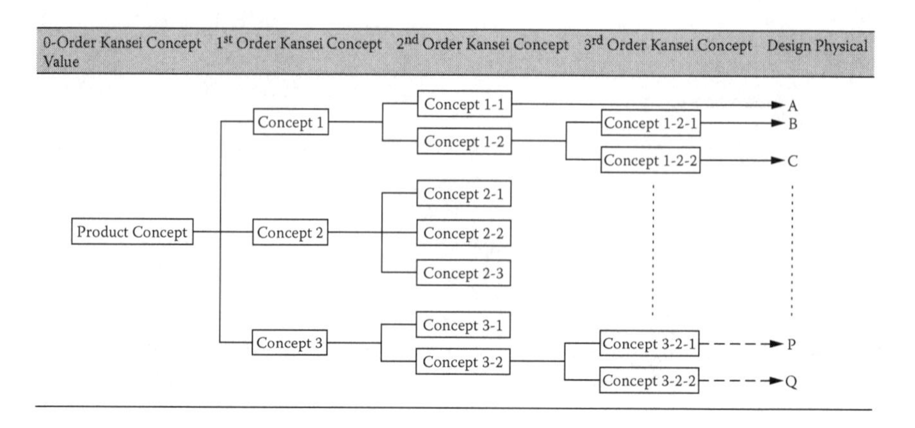

FIGURE 3.1
Conceptual map of Kansei engineering method Type I.

well as other product trends for this target group. In Kansei engineering, the product concept is called *zero-order Kansei concept.*

In this step, if we just pick a specific product (for example, a compact car for women) taking a wild guess for the target group, there is a chance we will fail.

3.1.1.3 Step 3: Breaking Down the Product Concept

Since the objective of Kansei engineering is to create a specific product that matches the human's Kansei, with just a product concept, there is no reference to the size, the kind of function, or the color of the product. So, as shown in Figure 3.2, the product concept is broken down into several levels until the design's physical characteristics can be properly assigned.

First, try to express what the content of the product concept is all about in another concept. There are several subconcepts that can be used to break down the product concept. I will try to describe all of them in detail. The first subconcept in the product concept level is called *first-order Kansei concept.*

In this step, the KJ Method can be applied. For example, let's say the zero-order Kansei concept is *people friendly.* Write down all the words that describe this concept on individual KJ cards. Group words according to similarity, and give a title to each group. As can be seen in Figure 3.3, two groups are created for the *people friendly* concept in the first-order Kansei concept, and let's assume the groups are titled *friendly to all* and *simple,* respectively.

At this stage, there is still no clue about what the *people friendly* product would look like. Therefore, the cards in each group are categorized and deployed to further lower-order concepts that describe each category. In the case of the KJ Method, the selected cards are taken out and put as a title for

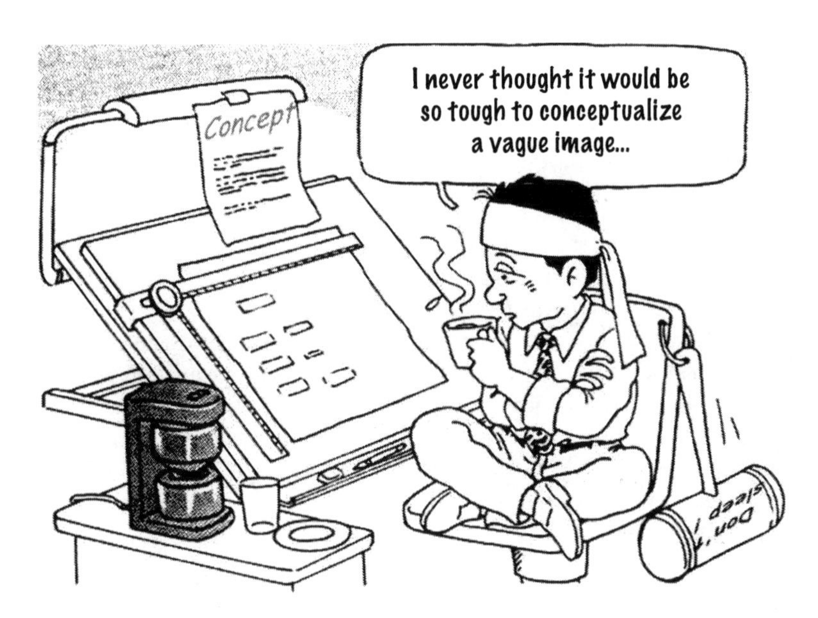

FIGURE 3.2
The breaking down of product concept.

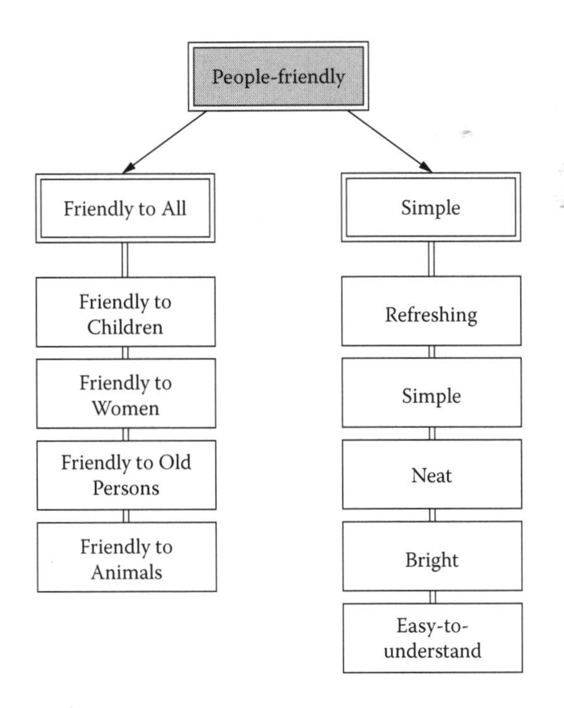

FIGURE 3.3
Kansei engineering Type I using the KJ Method for the concept *people friendly*.

each group, and words that are associated with them will be included in the group. It will result in something like what is shown in Figure 3.3.

3.1.1.4 Step 4: Deployment to Physical Design Characteristics

Compared with Figure 3.3, the concept has become more detailed and easier to understand in Figure 3.4. In this example, the *friendly to all* concept has been deployed up to the third-order Kansei concept, while the *simple* concept is up to the fourth-order Kansei concept.

When we reach this stage, we are able to figure out the weight and shape of the new product, such as lightweight, easy to carry, and so on. We can also relate the concept to technical words to figure out what kind of automation it will do, by keywords such as *function is simple* or *automation.* From keywords such as *simple appearance, lightweight,* and *easy to carry,* we can figure out the appearance design, while from *bright tone* we can derive the color for the appearance design.

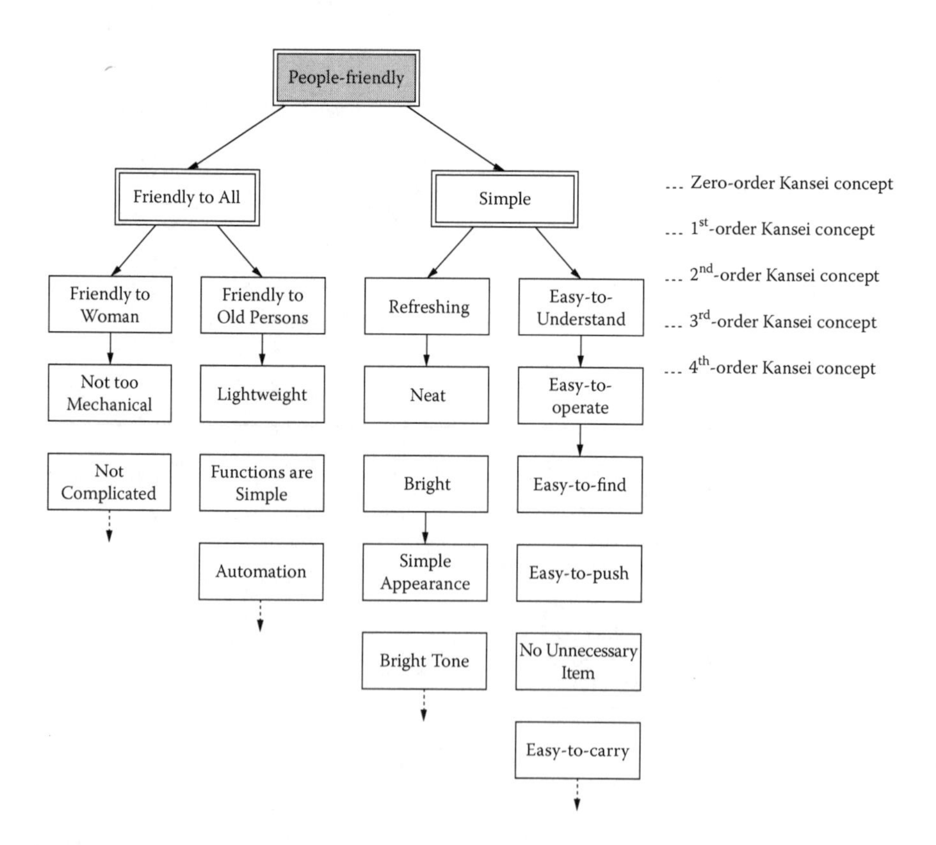

FIGURE 3.4
Expansion of Figure 3.3 to third- and fourth-order Kansei concept.

As described above, when we break down further and further the product concept, which at first was very general and vague, into lower-order concepts, we will reach a stage where the physical design characteristics will appear without us even realizing it. At this point, we can go on to decide the physical characteristics such as size, appearance design, color, and function.

3.1.1.5 Step 5: Translation to Technical Specifications

Even though the physical characteristics have been decided, that does not mean that we are ready to design the new product. At this stage we have only identified its characteristics. Now, we need to translate these characteristics into technical specifications. For example, in order to realize the *automation* concept that appeared in Step 4, we need to develop a new automation mechanism based on the past models and the next new level of function. In terms of design as well, we have to determine through trials and experiments, for example, the size and kind of function (SF property) a button should have in order to achieve the *easy-to-understand* and *easy-to-push* concept.

This is how using the Type I method helps organize the lower-order concepts into details and reaches the level of the physical characteristics, and then translates those into the technical specification level. Many technical steps are performed, and these result in the invention of a new design.

3.1.2 Application in Passenger Car Design

The most appropriate and easy-to-understand example for application of Kansei engineering Type I is the Mazda Eunos Roadster (MX5).

This car had been conceptualized with careful consideration as a sporty car for young people based on marketing surveys and several previous experiences. This was the decision from a target market point of view. Then, after detailed investigation of the lifestyle of young people (in their 20s) and long discussions, the zero-order Kansei concept was decided as "the rider and horse as one" (Jinba Ittai). The meaning of this concept is cited from the Eunos Roadster's article that can be found in Box 3.1.

According to the article, "the rider and horse as one" expresses the feeling of the driver and the car united as one, which is explained as "the rider (driver) who sets passion and energy in order to generate maximum power of the horse (car). And the horse instantly senses and reacts to the rider's slightest movement, unites with the rider and runs throughout the land as intended. It's the feeling of oneness between the rider and his horse…". The process of associating this Kansei to the actual design, if illustrated using Kansei engineering Type I, will look like Figure 3.5.

The zero-order Kansei concept, "the rider and horse as one," was broken down to *tight feeling, direct feeling, running feeling,* and *communication* as the first-order Kansei concept. However, even after breaking down to these four,

BOX 3.1 THE STORY OF *JINBA ITTAI*

On the business card of the project general manager who was in charge of the Eunos Roadster's development, *jinba ittai* (人馬一体) (the rider and horse as one) has been printed in vigorous brushstrokes. It was not *jinsha* (人車) (the driver and car). Actually, it was *jinba* (人馬), that is, the rider (driver) who unsparingly pours out passion and effort in order to bring out the maximum power of the horse. And the horse, which instantaneously senses and reacts to the rider's slightest movement, unites with the rider and courses through the land as he likes. It is the feeling of oneness between the rider and his horse that is the strongest bond between the driver and sports car, and this is symbolized with the *jinba ittai* phrase. For this project, all the key members of the project staff stayed together day and night in a loft on a river inside the factory premises (they affectionately call the loft Riverside Hotel). They did that to really feel the *jinba ittai* feeling, which could not be fully expressed in specifications: the naturally aspirated dual overhead cam engine that does not have the supercharging mechanism, the frame that integrates the drive train system using the power plant frame, the lightweight and underslung layout that keeps the Yaw-inertia moment at the minimum, the highly rigid open body, and the tight two-seaters. All these are the result of the pursuit of enjoyment in driving by bringing out the car performance to its fullest through the absolute specs and high mechanism. It is not *jinsha ittai*, but *jinba ittai*. Grab the Eunos Roadster's steering wheel and feel the comfort of the bond.

the physical characteristics still could not be reached, so the second-order Kansei concept was further broken down.

Here, the *tight feeling* means the driver and car are in close contact. This Kansei cannot be experienced in a full-sized or compact car. In other words, it is equivalent to a Kansei called *appropriate narrowness*, and it is decided that the car should be about 4 m long. The final physical length is 3.98 m, as a result of considerations of other functions that were performed later. Then, results from the second-order Kansei show that the car will be cramped if four seats are mounted into it. Thus, it was decided it would be a two-seater car.

As described above, in the process of breaking down a vague Kansei to a few levels, a specific physical value and characteristics could be identified. The next step is to bring these characteristics into the field of automotive engineering and convert them into the appropriate design element.

For the Eunos Roadster, a lot of interesting developments from the Kansei engineering point of view were performed and patented. One example is the length of the shift lever. In Figure 3.5, the *direct feeling* was broken down to *driver wish* and *maneuvering feeling*.

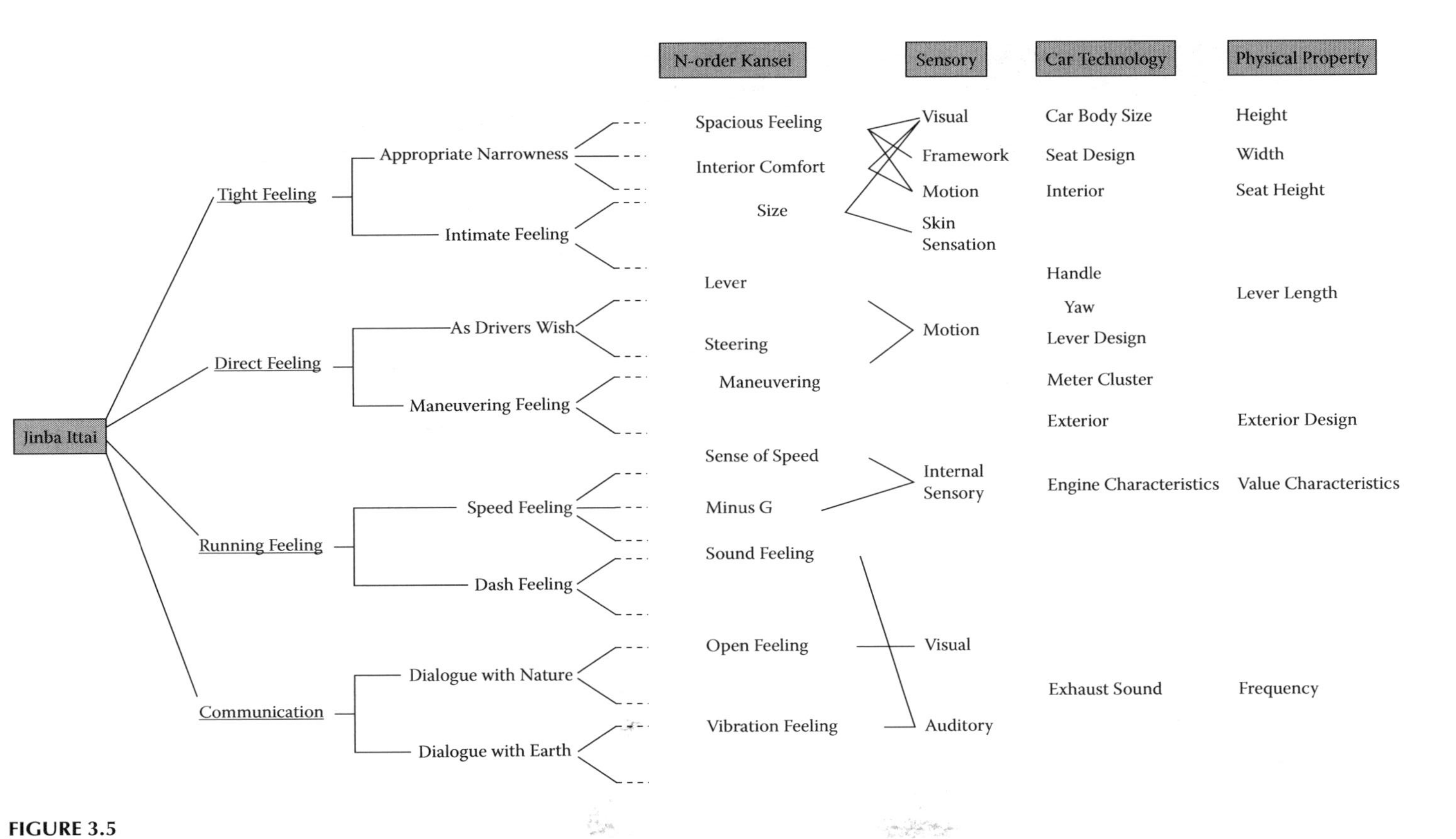

FIGURE 3.5
Translation map from Kansei for product's physical characteristics.

In order to realize these Kansei, one needs to know the optimum length of the shift lever (gear lever). Designers prepared shift levers of a few lengths and performed some experiments. The Kansei evaluation by the research workers showed that 9.5 cm from the joint is the most appropriate length, and so they implemented it.

Then, for the *dash feeling*, they succeeded in emphasizing the feeling by making use of the exhaust sound. They used a method in which the actual exhaust sound was recorded and experimented with, but the experiment could not be controlled and the factor value was also unknown. Since the vehicle sound can be processed using rotation and component order ratio, where the frequency that is equivalent to the engine's rpm is taken as the base frequency, they developed a vehicle sound simulation system using this characteristic.

They performed simulation experiments on the system, and with sensory tests using their research workers, they achieved the most suitable sound character for the *dash feeling*. In a 4-cycle, 4-cylinder engine, combustion occurs twice per engine rotation. Therefore, the even degree component of engine rotation will be the principal component of the vehicle sound. Odd degree component and 0.5 (half) degree component sounds occur from unbalanced factors of the engine system. By right, these sounds should be reduced, but from the sensory test result, if these sounds are reduced, it will result in *something missing* Kansei. In fact, they found that a high half degree component will produce *impressive* Kansei. Eunos Roadster's muffler was designed by heeding this kind of Kansei sound experiment.

3.1.3 Application in Brassiere Development

The top lingerie maker, Wacoal, had produced many bestsellers in its brassiere products line, especially the line called Good Up Bra. Within 2 years after it was developed, the Good Up Bra recorded a huge sale of 3 million pieces. This is the highest sale volume ever achieved by Wacoal, as I mentioned in Chapter 2. The development of the Good Up Bra was supported by many factors including Wacoal's vast experience, ideas, and technologies. For example, Wacoal has accumulated databases related to anthropometric and biological human engineering, which are the foundations for lingerie making. Even if we look at many of its successful products, we can imagine the richness of Kansei related to women in their R&D staffs. Furthermore, Wacoal's endless ideas and efforts in the field of technology development have also resulted in successes such as the Kokochi E brassiere with the introduction of the shape-memory alloy and the Three Quarter Cup with the utilization of amorphous fiber. The Good Up Bra is the pearl of wisdom of Wacoal's R&D staff and designers, but the Kansei engineering way of generating ideas played a role in its success. Let's examine the product development process from the viewpoint of Kansei engineering Type I.

When Wacoal surveyed 2000 women who participated in its monitor program about the qualities of a brassiere that shapes beauty, they received many responses. The top two answers on the questionnaire were:

1. Both breasts should settle within the chest width.
2. Nipples should be in parallel and face upward evenly.

This is equivalent to the process of discovering the physical characteristics for the Kansei concept of *becoming beautiful*, which was done by summarizing the opinions of women gathered through surveys. The next stage would be realizing these two conditions using material and design development, in other words, the expertise of Wacoal's technology.

As a result of their investigation, they successfully developed the refined amorphous fiber that is strong yet soft to match with the sense of soft and sensitive breasts. They did this by experimenting with the existing amorphous fiber, and by completing the design of the cup shape that pushes the breasts up, and which creates a flattering impression among women with the *push in and up* concept.

Similarly, in the case of the Good Up Bra's development, the Kansei engineering Type I procedure was performed, translating the zero-order Kansei of the *beauty shape* concept into physical characteristics, and then into technical specifications. The difference between wearing the normal brassiere and this new product determined by Moire measurement is shown in Figure 3.6. Moire measurement is a method in which the object's three-dimensional characteristic is illustrated in a pattern of contoured lines by shedding a light on it. With it we can clearly observe the characteristic of the shape. If we look at Figure 3.6, in the case of a normal brassiere, the breasts are facing toward the outer sides of the chest. In the case of the new product, we can see how they are drawn inside within the chest width. In addition, we can see that both breasts are facing forward and slightly upward. The shape of the breasts is round and firm, in keeping with an image of elegant beauty.

3.2 Kansei Engineering Type II

3.2.1 Overview of Type II

The sequence for the Kansei engineering Type I starts with breaking down the concept and reflecting it into a design's physical characteristics. Then comes the translation of physical characteristics into technical specifications. Kansei engineering Type II is similar to Type I, in that they both start from a Kansei concept. However, it differs from Type I in that the Kansei concept

When Wearing the New Product

When Wearing the Wireless Brassiere

FIGURE 3.6
Difference when wearing the new product and the wireless brassiere by Moire measurement.

is converted into physical characteristics (here, it is called *design requirement* or *design element*) using a method called Kansei engineering technique (or *translation technique*).

Look at Figure 3.7. The Kansei engineering Type II is a technique of translating the image or Kansei of a product that consumers hold in their minds into tangible product design elements. For example, imagine that we are now going to design the shift lever of a passenger car. If the user wants a sporty car with a shift lever that gives the *running feeling* and the feeling of *self-control,* we have to identify the physical characteristics that can simultaneously

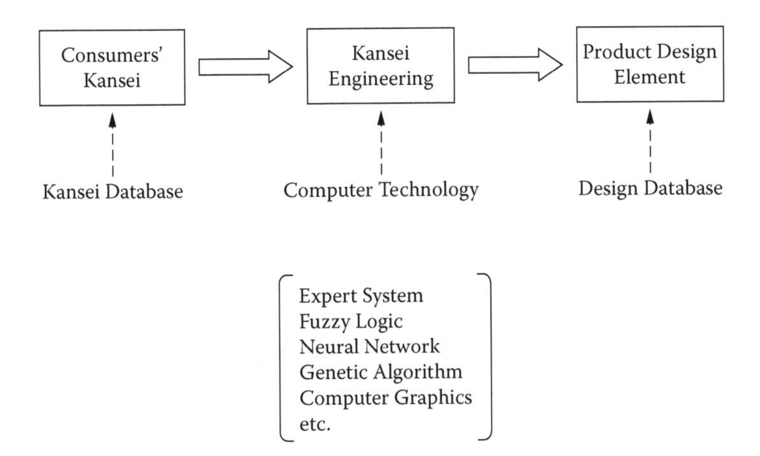

FIGURE 3.7
Translation process in Kansei engineering Type II.

realize both the Kansei of *running feeling* and of *self-control*. The Kansei engineering Type II is a method that has a database of consumers' Kansei and translates both relations with physical characteristics. It is a process to discover (1) length, (2) knob size, (3) knob shape, (4) range of shifting, (5) shift torque, and other characteristics of a shift lever that best fit both Kansei. If the translation technique of Type II concludes the shift lever length must be 9.5 cm, it means this Type II translation technique produces the same conclusion as Type I, which was used in developing the Eunos Roadster.

The essential components for the Kansei engineering Type II as shown in Figure 3.7 are the following three points:

1. All the Kansei held in the consumers' minds regarding the object must be covered, and a database that consists of all those Kansei should be established.
2. A design database that consists of all the design specifications related to the artifact should be established.
3. An inference function capable of linking the Kansei and design specification is required.

If the Consumers' Kansei at the left-hand side of Figure 3.7 can respond to point 1, the Product Design Element at the right side of Figure 3.7 can respond to point 2 , and the Kansei Engineering in the middle of Figure 3.7 can respond to point 3, then the Kansei engineering Type II is successful. In other words, if we establish points 1–3 and associate them to one another, we have successfully built the Type II system.

In order to establish such a system, that is, a kind of intelligent system, Kansei engineering Type II and a computer are employed. In establishing this intelligent system, techniques related to computer science such as Expert

System, Neural Network, and Genetic Algorithm (GA) are used. If necessary, Fuzzy Logic might also be used. At the moment, Kansei engineering that has the highest possibility to be deployed into computer technology is Kansei engineering Type II.

In Kansei engineering Type II, a lot of words that are related to consumers' Kansei will be collected, and the meanings of these words will be investigated. In order to achieve this, statistical analyses are performed. On the other hand, the design elements of the targeted product must be broken down, and these elements need to be arranged in the form of item/category classification. Based on experimental evaluation, survey, or designers' skill, we need to produce some kind of modeling to link the two. When these three points are integrated, the Type II technique is complete. I will describe this method in further detail in Chapter 4.

3.2.2 Application of Type II

In Chapter 2 we described a case study on a computer system called HULIS. This is a typical example of Kansei engineering Type II, and it is the earliest form of its implementation. HULIS is the acronym for *human living system*. It is a Kansei engineering system that was developed imagining a scenario of a consumer who wants to build a house and visits a construction firm to find the house that matches his Kansei. HULIS is divided into seven options: appearance, structure, entrance, Japanese-style room, Western-style room, kitchen, and bathroom. The system begins with the user input of the image or Kansei that he has in mind when the computer asked "How would you like it to be?" for each option. Besides being able to understand the word (Kansei word) input by the user, HULIS at the same time can also process correlated words because it can accept up to 10 words per input. The Kansei word database covers all the words typically used in the construction industry by the client and the architect. The database is also capable of understanding the structure of words using factor analysis related to house design.

Regarding the design elements of a house, data that were broken down to many specifications are stored in a computer. See the entrance design in Figure 2.3 as an example. The design shows a double-panel door, but may get more than 100 conditions if we further break down conditions such as (1) Single door, single panel, or double panel? (2) With or without fanlight? (3) An arch instead of fanlight? (4) How many glazing bars? (5) How many vertical bolts? (6) With or without lattice, and style of lattice? (7) Color of the door? (8) Width and shape of the entrance? This is called the item/category classification, which groups the design elements. The database is prepared using these design elements identified from all sections from external appearance to bathroom designs.

If the consumer wants to have a *gorgeous* entrance door, the system will generate a door design that fits the *gorgeous* Kansei and will show it on the screen. In this case, the computer just needs to interpret and suggest the

Kansei *gorgeous* with the kind of door frame. Is a fanlight necessary? How many glazing bars and vertical bolts will do? Is a grid necessary, and what kind of grid is suitable? Also, what is the suitable color for the door? In HULIS, these relationships are obtained from an experiment that involves the experts (architects) as test subjects, and are stored in the computer as a knowledge database. There is also a method to create a database from evaluation data from general users.

A system that was made of these three conditions is HULIS. For the seven options, when the desired Kansei words are input for each option, illustrations like Figures 2.3 and 2.4 will appear on the screen.

A space diagnostic system (IKDES) as in Figure 2.1 was also created using the same procedure as in Figure 3.7. In this case, Kansei words related to a car interior were collected and stored in a database. In order to apply only *spacious feeling* and *narrowness feeling* from the huge number of Kansei words, the system was limited according to its application purpose. On the other hand, detailed analysis was performed with regard to car interior designs, which were broken down to about 200 design elements and stored in a database. They form part of a diagnostic system as illustrated in the system diagrams (Figure 3.8).

Evaluation experiments were performed using the internal designers as test subjects. Actual passenger cars were parked at one place, and the subjects were asked to get into those cars and evaluate the interior designs. A multivariate analysis was performed with the evaluation results. Statistical

FIGURE 3.8
Illustration of the use of IKDES.

links between Kansei words and design elements were then stored into a knowledge database.

Up to this point, IKDES and HULIS work almost the same. The unique point for IKDES is that when a designer inputs a value that he wants to design into a screen that appears on the display, the system will immediately perform a multivariate analysis based on the value and diagnose whether it is appropriate or not in terms of Kansei. IKDES has evolved from a computer system into a designer support system, which is easier for designers to use than HULIS.

3.2.3 Consumer Decision-Making System and Designer Support System

The Kansei engineering system can be roughly divided into two systems: the consumer decision-making system and the designer support system. The consumer decision-making system helps consumers make decisions when selecting a product that matches their Kansei. HULIS is the typical example of it. The designer support system is a system that serves as a reference to a designer who is developing a new product by making a design that matches the targeted product Kansei, and at the same time, the system will diagnose how close the design is to the desired Kansei. Both systems have their own roles, and they were created to serve those roles.

A new Kansei engineering system is then formulated. It is called the hybrid Kansei engineering system. I will explain this in detail later. In brief, it is a system that aims to efficiently support the designer's new product design from the perspectives of the system that translates from Kansei to product design elements and the execution of Kansei diagnosis from design input. As illustrated in Figure 3.9, the flow for the former is called the forward Kansei engineering system, and the flow for the later is called the backward Kansei engineering system. The combination is called the hybrid Kansei engineering system. There are some completed systems for certain

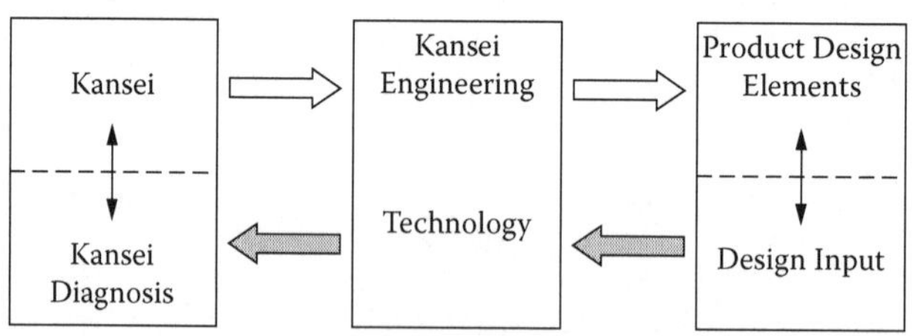

FIGURE 3.9
Schematic diagram of hybrid Kansei engineering system.

fields. By introducing this system, new product development will be much more efficient and the lead time will be shortened.

3.3 Kansei Engineering Type III

The process that starts with a Kansei study and then is reflected in physical design characteristics is also the same in Kansei engineering Type III. The difference is that in Type III, a mathematical model is mediated, and the relations from input to output (physical characteristics) are established by finding the coefficient value.

An easy-to-understand example is the Kansei engineering of Word Sound Image, which has already been explained in Chapter 2.

3.3.1 Hardware That Generates Sound

When mammals such as dogs and monkeys produce sounds to express certain emotions, they will produce sounds that are common to their species. In addition, the shape of the mouth opening is relatively fixed for each emotion. For example, when a dog wants to express affection, its breath will flow out through its nose and it gives a "koon, koon" sound. When threatening others, it will show its teeth, breathe heavily through its mouth, and bark. Similarly, there is a discipline that studies human voice production called *articulatory phonetics.*

An anatomical drawing of a human palate is shown in Figure 3.10. The voice is changed by sending breath to the nasal chamber and to the oral cavity, which is achieved by opening and closing the pharynx. The voice will also differ when the oral cavity is widened or narrowed. Putting the upper and lower lips together or setting them apart, and clamping teeth together or setting them apart also will result in different voice sounds. In other words, humans express different emotions using different voices, by controlling the shape of the anatomical hardware from the throat to the mouth and nose.

If this is arranged according to the notational system in articulatory phonetics, the anatomical mouth movement can be shown with phonetic symbols as in Table 3.1. By using this, except for vowels, *n,* and *sokuon* (little *tsu*), the Japanese language can be roughly categorized according to manners of articulation and the combined vowels. Articulation can be categorized according to manners of articulation and places of articulation (anatomical places that produce voice) (Table 3.2). If we add the vocal cord vibration (voiced and voiceless) categories such as *pa* row and *ba* row, *ta* row and *da* row, the categorization is almost complete. In other words, all consonants in

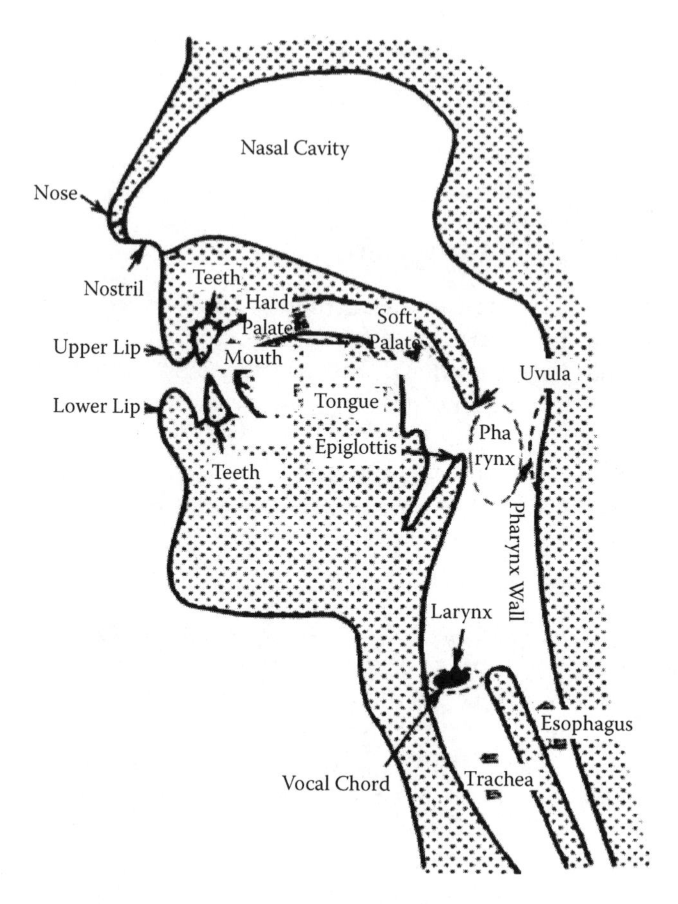

FIGURE 3.10
The anatomical chart regarding articulation.

the Japanese language can be categorized into Tables 3.2 and 3.3—manners of articulation and places of articulation.

Vowels are generated from glottal vibration, and they are mainly determined by the position and shape of the tongue. Vowels can be categorized as shown in Table 3.4. With Tables 3.2, 3.3, and 3.4, we can express Japanese language articulation and understand its characteristics.

3.3.2 Survey on Word's Semantic Differential

In the preliminary experiment, 40 pairs of adjectives (e.g., warm–cold) that were considered as expressive words' Kansei were selected and set into SD Scale format. After evaluating a few Japanese words using those adjective pairs, we performed factor analysis and selected 20 pairs of adjectives that are most suitable to describe the words' Kansei. They are shown in Table 3.5.

TABLE 3.1

Phonetic Symbols

		Bilabial	(Denti-) Alveolar	Post-Alveolar	Palatal	Velar Uvular	Glottal
Consonants	Nasal	m		n		N	
	Plosive	p, b	t, d		ç	k, g	
	Fricative	Φ	s, z	ʃʒ	c		h
	Approximant				j	ɯ	
	Flap		ts, dz	tʃ dʒ			
	Affricate		ɾ				

Nasal	*ma* row (m) *na* row (n) *n* (N)
Plosive	*pa* row, *ba* row (p, b) *ta* row, *da* row (t, d) ki (ç) ka row, ga row (k, g)
Fricative	fu (Φ) sa, za (s, z) shi, ji (ʃ) (ʒ) hi (c) ha, he, ho (h)
Approximant	wa (ɯ) ya row (j)
Flap	tsu, dzu (ts, dz) chi, ji (tʃ, dʒ)
Affricate	ra row (ɾ)

TABLE 3.2

Places of Articulation Categories

Description		Places of Articulation Pronunciation Method
Bilabial	pa row, ba row	Close both lips, then release.
	ma row	With both lips closed, breathe out through nasal cavity.
	fu	Narrow the opening of lips and create breath friction.
(Denti-) alveolar	ta row, da row	Release breath with tongue tip and gum ridge, then release.
	ra row	Release breath with tongue tip and gum ridge, breathe through nose.
	na row	
	sa row, za row	Narrow the gap between tongue tip and gum ridge, create friction.
	tsu, tzu	
Palatal Post-alveolar	shi, ji	Create breath friction using edges of tongue placed between gum ridge and palate.
Palatal	ki	Stop breath with tongue's front surface and palate, then release.
	hi	Create breath friction by narrowing the gap between tongue's front surface and palate.
	ya row	Narrow the gap between tongue's front surface and palate to an extent that does not cause a breath friction.
Velar	ka row, ga row	Stop breath with tongue's rear surface and soft palate, then release.
	wa	Create breath friction using tongue's base and soft palate.
Nasal uvular	n	Stop breath with tongue's rear surface and soft palate, breathe through nose.
Glottal	ha, he, ho	Create breath friction by narrowing glottal without vibrating the vocal folds.

TABLE 3.3

Manners of Articulation Categories

		Manners of Articulation
Description		**Pronunciation Method**
Nasal	ma row, na row	Release breath at certain location in oral cavity or at throat, and flow it out through nasal cavity.
Plosive	pa row, ba row ta row, da row ka row, ga row	Stop breath completely at certain location in oral cavity or at throat, then release it in a breath.
Flap	tsu, tzu, chi, ji	Stop breath briefly, and then release it gradually.
Fricative	sa row, za row ha row	Generated from the friction of breath with its surrounding when it flows through a very narrow passage.
Approximant	ya row, wa	Generated by obstructing breath to an extent that does not cause a friction, with lips rounded.
Affricate	ra row	Narrow the oral cavity to an extent that does not cause a breath friction.

TABLE 3.4

Japanese Vowel Categories

	Palate	Center of Tongue	Soft Palate
High	i		w
Middle	e		0
Low		a	

In the first experiment, we randomly arranged 68 characters that do not contain *sokuon* (little *tsu*), *yōon* (added y sound), and *chōon* (long vowel sound) and asked the subjects to evaluate each of them using the aforementioned 20 Kansei words. The evaluation scale was a five-level rating, and the test subjects were 28 university students of both genders.

When we translate the characteristic of each character categorized from the standpoint of phonology with the evaluation of the 20 pairs of adjectives, we can discover the Kansei for each of them. With that, we concluded that words' Kansei can be identified with the manner and place of articulation. Using Tables 3.2, 3.3, and 3.4, we created the item/category classification to categorize the 68 words. This is shown in the leftmost column of Table 3.6. Here, we considered the vowels' row as a separate indicator and included it as one of the item/category group.

As shown in Table 3.6, the 68 characters are categorized into row, manner of articulation, and place of articulation. Since the Kansei for *dakuon* (muddy sounds) is different from *seion* (clear sounds), these are categorized separately. This item/category classification was combined with the data evaluated using the aforementioned 20 adjective pairs, and was analyzed using the Quantification Theory Type I. Part of the statistical analysis result (only

TABLE 3.5

Adjective Pairs Used in Evaluation Experiment

1	Soft	Hard
2	Bright	Dark
3	Broad	Narrow
4	Unique	General
5	Expansive	Unexpansive
6	Heavy	Light
7	Refreshing	Old
8	Unambiguous	Ambiguous
9	Simple	Complicated
10	Glamorous	Unglamorous
11	Warm	Cold
12	Individual	Common
13	Have uplifting feeling	No uplifting feeling
14	Nice ring	Ill-sounding
15	Roundish	Squarish
16	Gentle	Unkind
17	Masculine	Feminine
18	Have sense of flowing	No sense of flowing
19	Sharp	Dull
20	Powerful	Powerless

for the warm—cold Kansei word) is included in Table 3.6. We also obtained results similar to these for the other 19 Kansei words.

The following explains how to read Table 3.6:

1. We have obtained the male data, female data, and overall data from a combination of both data. We can understand the differences between genders by looking at respective columns.
2. Since the multiple correlation coefficients for both are above 0.7, these results can be considered reliable.
3. As for the row, we can see that the *a*-row and *o*-row are minus for both genders, and both rows have warm Kansei. Besides, the *u*-row and *n* were also perceived as warm Kansei by females. On the other hand, the *i*-row and *e*-row were perceived as cold Kansei by both genders, while males felt the *n* was very cold (Figure 3.11).
4. In terms of the manner of articulation, both genders recognized the nasal (*ma* row and *na* row, which come from nasal cavity) and affricate (*ra* row, which is pronounced with the oral cavity narrowed to some degree) sounds like warm Kansei. Fricative (*sa* row and *za* row, which are generated when the breath rubs against its surroundings while flowing out) and flap (*tsu*, *tzu*, *chi*, and *ji*, which are pronounced

TABLE 3.6
Analysis Results for "Warm—Cold"

Adjectives: Warm—Cold				
***Seion* (Clear Sounds)**		**Overall**	**Male Data**	**Female Data**
Multiple Correlation Coefficient		0.7708	0.7646	0.7576
Item / Category		**Score**	**Score**	**Score**
Row	Partial correlation coefficient	0.5630	0.6040	0.618
	i-row	0.2422	0.1632	0.3865
	e-row	0.2185	0.1282	0.3881
	a-row	−0.2561	−0.3103	−0.1678
	o-row	−0.1590	−0.1111	−0.3003
	u-row	0.0291	0.1025	−0.1697
	n	−0.0307	0.8763	−0.4260
Manner of articulation	Partial correlation coefficient	0.5710	0.5620	0.5800
	Nasal	−0.2823	−0.2673	−0.3242
	Plosive	−0.0223	−0.0377	0.0123
	Fricative	0.2718	0.2177	0.3934
	Approximant and vowel	0.1072	0.1137	0.0924
	Flap	0.2951	0.3128	0.2551
	Affricate	−0.1753	−0.0871	−0.3737
Place of articulation	Partial correlation coefficient	0.6490	0.6480	0.5650
	Bilabial	−0.2248	−0.1770	−0.2988
	(Denti-) alveolar	0.1373	0.1260	0.1963
	Palatal – Postalveolar	0.0244	0.1604	−0.2482
	Palatal	−0.2127	−0.2605	−0.0713
	Velar	0.4592	0.5003	0.4003
	Glottal	−0.2575	−0.2460	−0.2499

by stopping the breath briefly and then releasing it) sounds were both perceived as cold Kansei.

5. In terms of the place of articulation, both genders felt that bilabial (*pa* row, *ba* row, and *ma* row) sounds and vowels were warm characters; on the other hand, denti-alveolar (*ta* row, *da* row, *sa* row, and *za* row) and velar (*ka* row and *ga* row) were perceived as cold sounds.

I will not explain the statistical result for *dakuon*. Even from Table 3.6 alone, we can discover a huge amount of information about the Kansei of word sound image.

To give an example in an actual case, in the words of a popular magazine among young females, *non-no* (ノンノ), we can see that the *n*, which is *o*-row and also a nasal sound, gives a warm image to females, and all three characters strongly give a warm feeling. Setting aside the articles and contents, we

TABLE 3.6 (continued)

Analysis Result for "Warm—Cold"

Adjectives: Warm—Cold				
***Dakuon* (Muddy Sounds)**		**Overall**	**Male Data**	**Female Data**
Multiple Correlation Coefficient		0.8555	0.8291	0.8259
Item / Category		**Score**	**Score**	**Score**
Row	Partial correlation coefficient	0.7200	0.6380	0.7870
	i-row	0.1722	0.2343	0.0324
	e-row	0.1134	0.0368	0.2856
	a-row	0.0750	0.0229	0.1921
	o-row	–0.0693	–0.0604	–0.0891
	u-row	–0.3310	–0.2333	–0.5509
Manner of articulation	Partial correlation coefficient	0.7890	0.7100	0.7870
	Plosive	–0.1386	–0.1049	–0.2145
	Fricative	0.3123	0.2254	0.5077
	Flap	0.5526	0.4615	0.7577
Place of articulation	Partial correlation coefficient	0.7500	0.7300	0.7020
	Bilabial	–0.1009	–0.1612	0.0349
	(Denti-) alveolar	–0.1383	–0.0832	–0.2623
	Palatal – Postalveolar	0.3222	0.2943	0.3849

FIGURE 3.11
Deriving the physical characteristics using a mathematical model.

can say that naming the magazine with a word that gives a good impression to females makes the magazine popular among young females. What I have explained up to this point is the result of research on the image of words using Kansei engineering Type II.

3.3.3 Word Sound Image Diagnostic System

We are now going to see an example of Kansei engineering Type III. Type III is a method that enables Kansei analysis to identify coefficient values by introducing some mathematical models.

In the aforementioned word image Kansei research, we deliberately excluded the *yōon* (e.g., *kya, shu*), *sokuon* (e.g., *kitt, shitt*), and *chōon* (e.g., *ki-, shi-*). All these words are considered single and independent letters. However, in reality, they contain characteristics such as dependency and sequence. Considering all these matters, we took the three *yōon, sokuon,* and *chōon* as operators that have little *ya*, little *tsu*, and "-" attached to them, prepared 80 samples, and performed Kansei evaluation. For dependency and sequence, including *n*, we created a dependency map for nouns that were randomly selected from a literary magazine, then performed Kansei evaluation on 66 samples that have the highest dependency.

Next, we created a computer system that will diagnose the word image Kansei for any kind of word input into it. In order to derive the word Kansei for a group of a few characters from the database of single character Kansei and dependency-weighted Kansei, we cannot simply add the result of the single character. Therefore, we introduced the fuzzy measure and integral model. Since it is difficult to explain in detail, I will skip the theoretical part. However, if we illustrate in a schematic diagram the process of determining the coefficient of Kansei diagnosis for a contiguous word from a single character using the fuzzy integral model, it will turn out as shown in Figure 3.12.

In this diagram, we enter the evaluation value of 20 pairs of Kansei words for the word that has no meaning into the big circle of Overall Evaluation Value at the right-hand side of the diagram. On the left-hand side, enter the value of the individual character for articulation, vowel, and operator, and calculate with the fuzzy integral model to get the weight value, *g*. Since this value will be the weight for this model, by using it, the model can diagnose the kind of Kansei for words.

A system diagram for the word sound evaluation system using a computer is shown in Figure 3.13. We have named it WIDIAS (word image diagnosis fuzzy expert system).

Ma-chi (マーチ) Kansei evaluation as shown in Figure 2.1 in Chapter 2 is one of the output results using the early WIDIAS system. It does not have data such as the *yōon, sokuon,* and *chōon*; however, it was capable of performing word Kansei evaluation quite precisely. The current WIDIAS-II is the most advanced. It includes not only conditions for the aforementioned *yōon,*

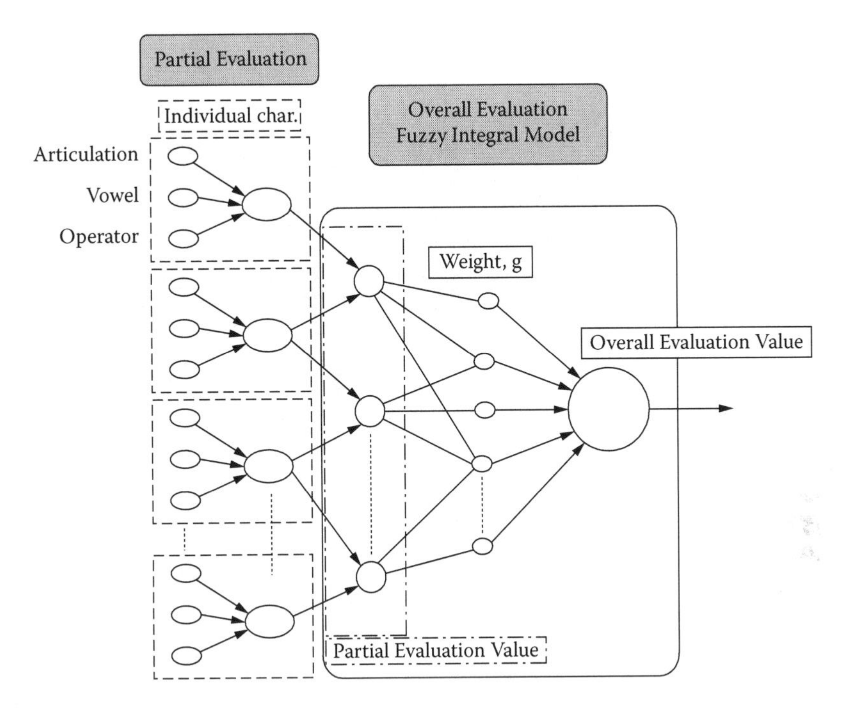

FIGURE 3.12
The sound of word evaluation model.

sokuon, and *chōon*, but also the knowledge of dependency and sequence. Its intelligence level has dramatically increased.

As an example, we input a magazine name that is popular among young females—*an-an* (アンアン) into the WIDIAS. The result is shown in Figure 3.14. If we look at it, the evaluation of artificial intelligence for the name *an-an* results in bright and refreshing; nice sound, gentle, and healthy; a sense of elegance and comfort; slightly feminine and weak personality, but also a sense of dull and less romantic. The artificial intelligence does not know that *an-an* is the name of a magazine purchased by females, but it gave a judgment that is very close to the image that we already have in our minds regarding the magazine. Therefore, we can say that the naming of this magazine is a success.

Next, let's work out a word that's a little bit scary—*jigoku* (じごく) (hell)—using WIDIAS. The result is shown in Figure 3.15. Of course, this system does not analyze the meaning of *jigoku* characters (地獄: the *kanji* for *jigoku*). It analyzes the Kansei of word pronunciation. The same as earlier, we divided it into right and left sides and began with the strongest Kansei.

This word gives such a bad impression that there were a very high number of the right side, minus Kansei, so high that I cannot include all of them here. In brief, the pronunciation for *jigoku* (じごく) turns out to be a word that is hard and dark, awkward, and not cute.

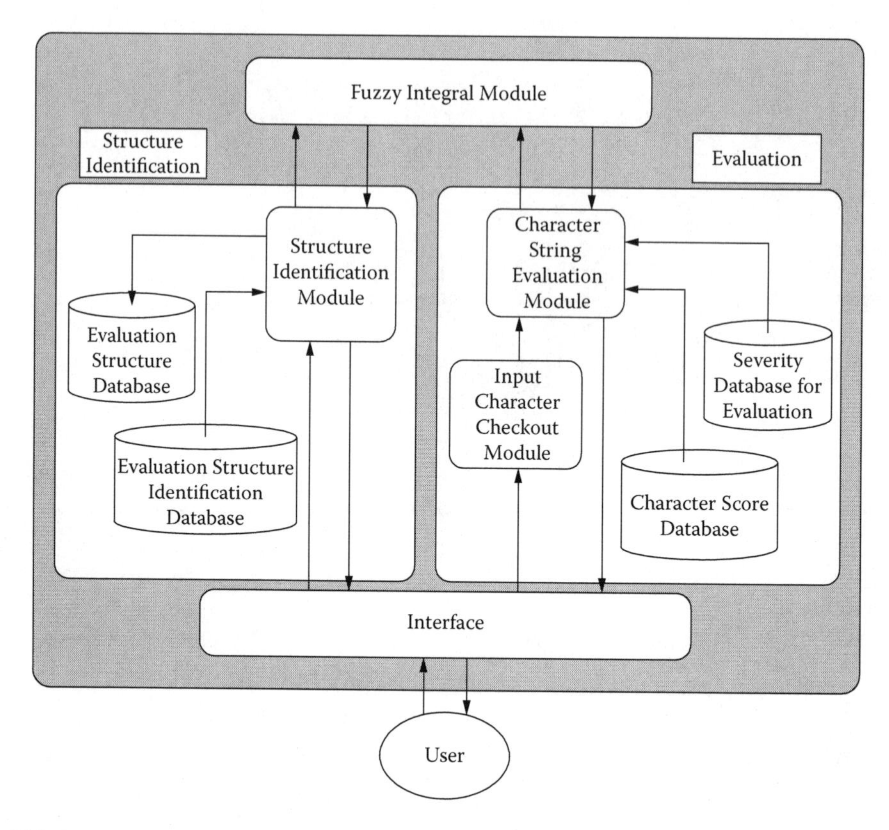

FIGURE 3.13
Schematic diagram for the WIDIAS system.

This WIDIAS-II can be used in selecting a name for a new product. For example, R&D people have completed the development of a new product based on some concept, and now they want to give it a name. Since the Kansei of the product concept has been fixed, the Kansei words that describe the meaning of the concept can be entered into WIDIAS. By doing so, WIDIAS will provide words that correspond to each Kansei word in a sequence that corresponds to the Kansei. Therefore, they can decide a name by combining a few characters from the top of the resulting list. Currently, anybody can acquire the WIDIAS-II from the company KDD Soken.

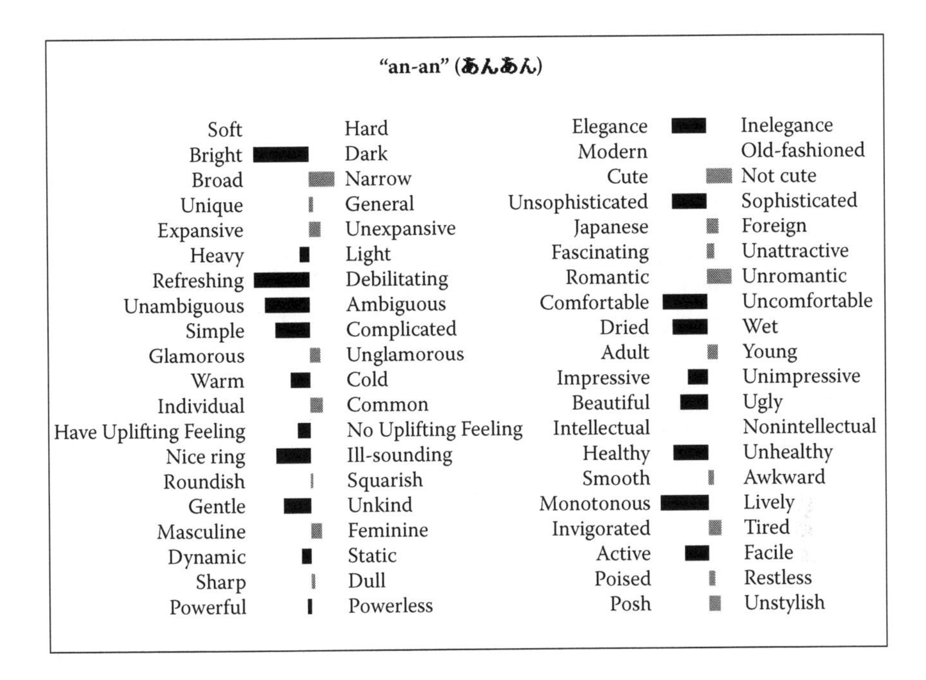

FIGURE 3.14
Kansei evaluation for *an-an*.

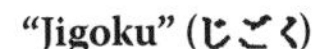

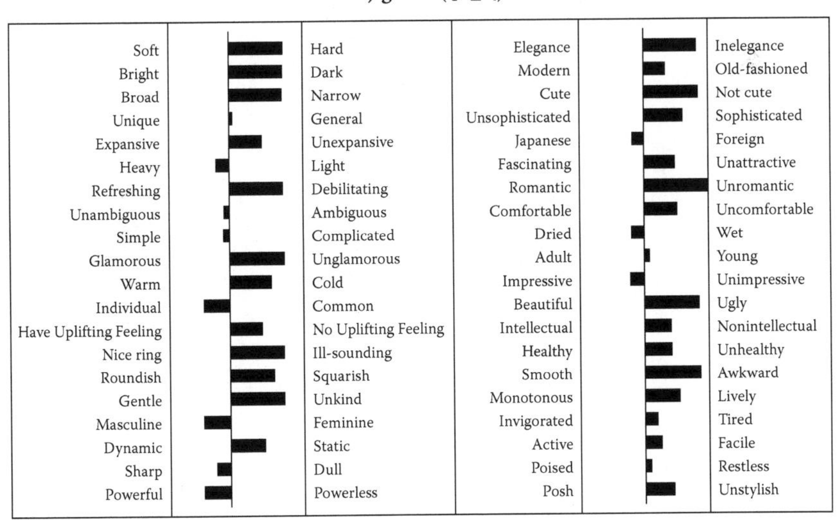

FIGURE 3.15
Kansei evaluation for *jigoku*.

4

Kansei Engineering Procedures: Kansei Engineering Type II

I have described the three types of Kansei engineering techniques—Type I, Type II, and Type III. I have explained in great detail the procedures for Type I. Since I have not detailed Type II, which is the most widely used, I will explain it in further detail here. As for Type III, except for the construction of mathematical models, it is similar to Type II.

If we model the procedure in conducting Kansei engineering Type II, we have something like Figure 4.1. Looking at the diagram, you might feel that it is complex, but actually it is not really that complicated. Anybody can do it almost effortlessly.

Anybody can build a Type II Kansei engineering system by following Steps 1 through 15. However, in this process, the most important things are (1) learn to see the subject design from the Kansei viewpoint, and (2) learn which points to focus on, that is, discern how people see those points. I will explain the procedure in sequence.

4.1 Selecting Survey Target

Kansei differs according to product, and some Kansei belong only to a particular product. When performing a Kansei engineering survey, especially for the purpose of new product development, it is recommended to limit to a specific product domain. Otherwise, products with different characteristics will mix in and create confusion in the analysis (Figure 4.2).

The target selection here does not refer to the industrial domain, such as automotive or home appliances. Take the automotive product as an example. There are different types, such as trucks, commercial vehicles, passenger cars, and RVs (recreational vehicles), according to their intended purposes. Thus, it is necessary to select the same type of product as the target from the viewpoint of product design.

Therefore, for a Kansei analysis related to passenger car design, it is crucial to narrow the scope of surveys and experiments only to passenger cars, without mixing them with trucks or other vehicles. However, passenger cars also range from light motor vehicles to ordinary motor vehicles, and the ordinary

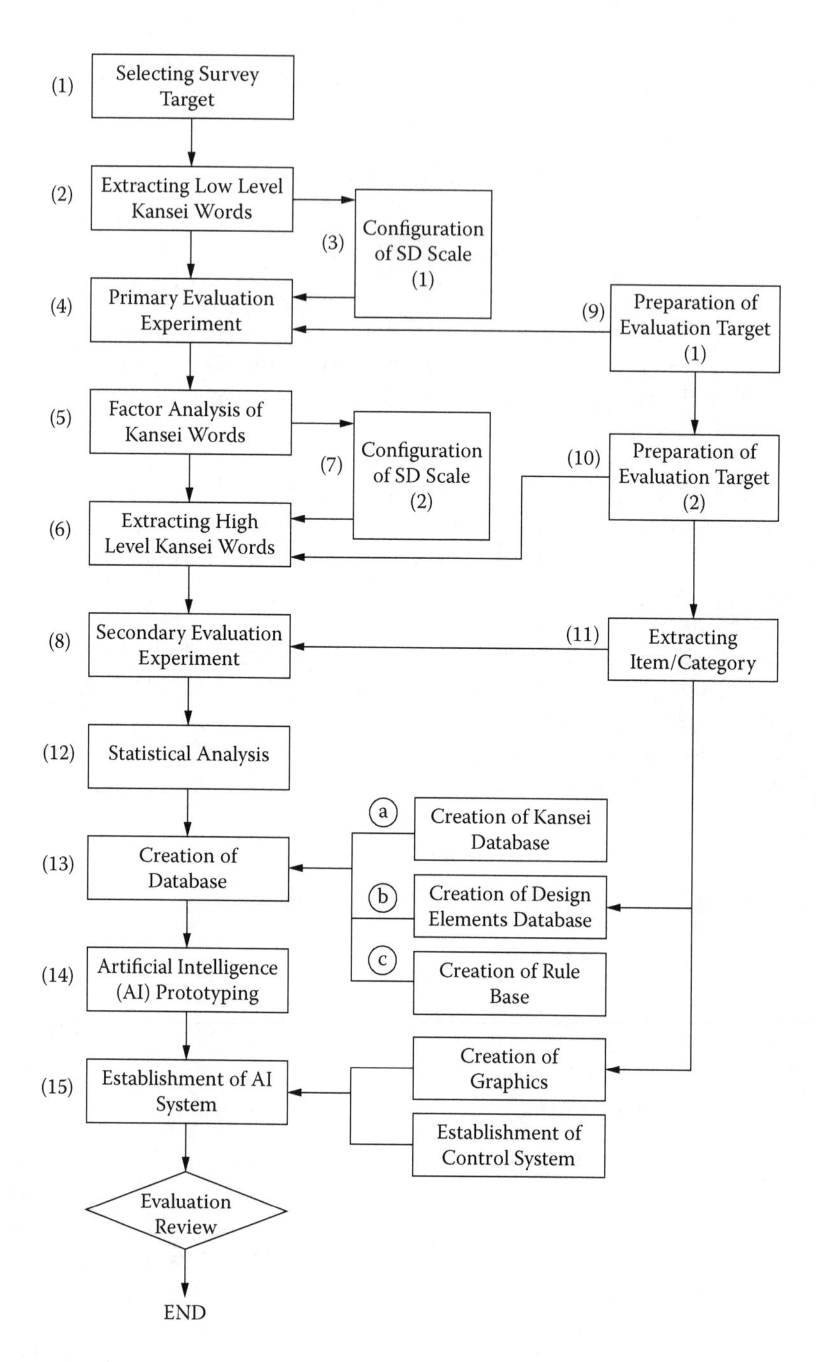

FIGURE 4.1
Kansei engineering Type II procedure.

FIGURE 4.2
Selecting survey targets.

vehicles too range in engine size from small 1-liter cars to large passenger cars of more than 3 liters.

There are also cases where we need to narrow down further depending on what we want to know. For example, if we want to know the Kansei relation with mini-car users, then we select just mini cars as the target. If we want to focus on the small-sized segment of ordinary vehicles, we limit the scope to passenger cars of 1 to 1.5 liters. Of course, if we want to grasp the Kansei characteristics for ordinary motor vehicles in general, we set general ordinary cars as the survey targets without including mini cars.

Under the home appliance product category, including general kinds of home appliance products as a survey target is a mistake. Some products are related to the kitchen, such as refrigerators, gas ranges, electric pots, and dishwashers; and others are related to audio-video equipment such as televisions, radios, and tape recorders. We will get better data if we limit the target to one of these items.

Furthermore, even for the passenger car, we can divide the category into exterior and interior. Even for the interior, depending on whether we want set the target as the steering wheel (handle) design only or not, there will be an infinite number of options.

It is important to clarify what we want to know and to try to keep the scope as narrow as possible.

4.2 Extracting Low-Level Kansei Words

After we selected the survey target, the next thing to do is to check the Kansei of the target. We need to know what kind of Kansei consumers feel toward a target product, but there is no method to directly measure the Kansei. There is only an indirect measurement, where we take some human responses as hints.

Please refer to Figure 4.3. What we want to measure is the Kansei, which is the underlying consumer attitude when purchasing a product. However, Kansei is by its nature ambiguous and without shape. Since Kansei cannot be measured directly, we have to employ some indirect methods by using external measurement tools.

One of the methods is physiological measurement. It is a method in which brain waves (EEG), electromyography (EMG), electrocardiogram (ECG), and other physiological ergonomic indicators are measured while a consumer is using or looking at the product. These measurement methods are not bad, and they are sometimes necessary. This kind of physiological measurement method is essential, especially in the investigations and experiments regarding *comfort*, which center around sensory functions. This method is necessary for cases in which the ergonomic considerations that clarify the physical characteristics in the Kansei engineering Type I are required. This kind of physiological ergonomic method is effective for product function surveys. In the field of sensory inspection, the measurement that uses sensory organs like this is called *Type I sensory inspection*, while the preference survey such as *like* or *dislike* is called *Type II sensory inspection*.

In Kansei engineering, it is the entire product or its component elements that give an impression to consumers and elicit the feeling. This is what we

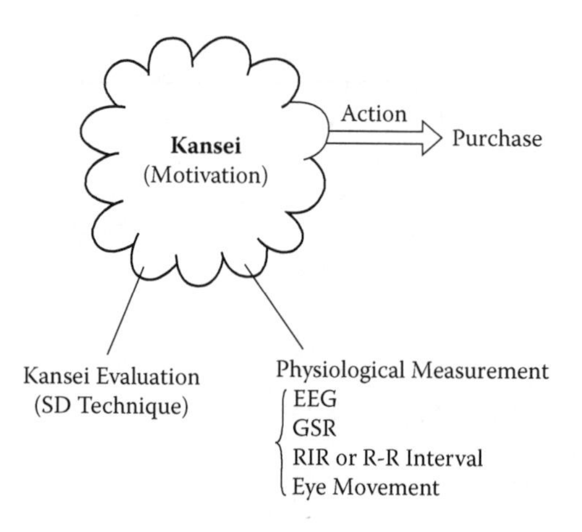

FIGURE 4.3
Indirect interpretation of Kansei.

call Kansei. That probably stimulates the brain to produce beta waves. It might stimulate the sympathetic nerves, making the heart beat fast. However, this does not give any clue regarding what part of the product is inducing what kind of feeling in the consumer.

Therefore, in Kansei engineering, an indirect measurement method has been adopted through words that are closest to Kansei expressions, like *elegant* or *flashy*. These are called Kansei words. Kansei words do not have to be limited to adjectives. Phrases like *the rider and horse as one* are also fine, but in most cases, consumers understand best if the words are expressed as adjectives. Therefore, in this book, Kansei words are all expressed as adjectives.

The process of collecting all the Kansei words related to the selected target starts here. To achieve this, we have the following methods:

1. Record dialogues between customers and marketing staff at the shops that sell the product. Record the dialogue between both parties using a hidden tape recorder, and then write down only the words that fall under the Kansei word category. The words collected in this manner are the most important.
2. Turn every single page of catalogs, related magazines, and pamphlets that describe the product, and list all the Kansei words. If the selected domain is a passenger car, then *Motor Fan* magazine and others fall under the category of related magazines. Collect every single Kansei word available.
3. Write down Kansei words from dictionaries such as the *Koujien* dictionary. If you cannot get the targeted number of Kansei words using Steps 1 and 2, extracting Kansei words from good dictionaries is a good idea. When pulling Kansei words from a dictionary, it is better to pick words related to Kansei regardless of their relevance to the product, since we will have the chance to arrange them later. It will be a versatile Kansei word collection and can be used for other products as well.
4. Ask the product designer to write down Kansei words related to the product. The designer will have certain favorite words, so make sure to collect them all.

It is recommended to collect about 600 to 800 Kansei words if possible, because if we have a big collection of Kansei words, it will be easier to narrow down the list to the most necessary ones. Even with this large collection, there is still a possibility that the necessary words will be left out. So, it is beneficial to collect words from the designers as well.

If we collect about 600 to 800 words and keep them as a general Kansei words list, add words related to other products, then arrange them into categories, it will be very convenient because we can use them anytime. If we input and keep them in a computer, they can be utilized in many convenient ways.

4.3 Construction of Semantic Differential Scale (Part 1)

The SD technique is a method known as *semantic differential,* formulated by Osgood and colleagues. Since there is no good equivalent meaning in Japanese, we called it the SD technique. Osgood and his associates were interested in the semantic space and had the idea to study words as a scale. They prepared many antonyms like *good—bad* or *elegant—crude,* surveyed the images held in humans' minds, and analyzed the data using factor analysis.

When selecting a word to study its image, Osgood and colleagues also prepared its antonym, for example, *elegant* and *crude.* This is because they thought that the semantic space of a word is made up of dimension(s) equivalent to the number of word(s) prepared. Additionally, the dimension for each single word is one straight line, and the human mind's image can be measured with a scale between the antonyms.

Furthermore, Osgood and colleagues thought words that normally express physical quantity for weight or brightness, like *light—heavy* or *warm—cold,* have the potential to be in a scale that can measure the human mind's image for any kind of subject.

Following the theory of Osgood's SD technique, Kansei engineering also holds a similar thinking that adjectives have the potential to be an image scale (a sensor that measures psychological emotions). However, we think that *elegant* and *crude* are not necessarily expressible using a straight line (a symmetric line that has a 180 degree angle), and in regard to product design, a measuring method using a sensor of *crude* is inappropriate. Words like *bad, crude,* or *dirty* are words expressing a lack of value; their negative side is not good in itself. These are not appropriate for evaluating products or comfort.

Therefore, in Kansei engineering, rather than antonyms like *elegant—crude,* a word that denies the other word is used, for example, *elegant—inelegant.*

In evaluating a subject using opposite words, there are concerns about how many levels should be used. There are a lot of options (see Box 4.1), such as (1) a 5-level rating method where the space between, for example, *beautiful* and *not beautiful* is divided into 5, and the evaluation is performed based on 5 levels; (2) a 7-level rating method where the space is divided into 7; (3) a 9-level rating method where the space is divided into 9. The maximum is 11 levels, and the minimum is 3.

BOX 4.1

① Beautiful |_|_|_|_|_| Not beautiful (5-level method)

② Beautiful |_|_|_|_|_|_|_| Not beautiful (7-level method)

③ Beautiful |_|_|_|_|_|_|_|_|_| Not beautiful (9-level method)

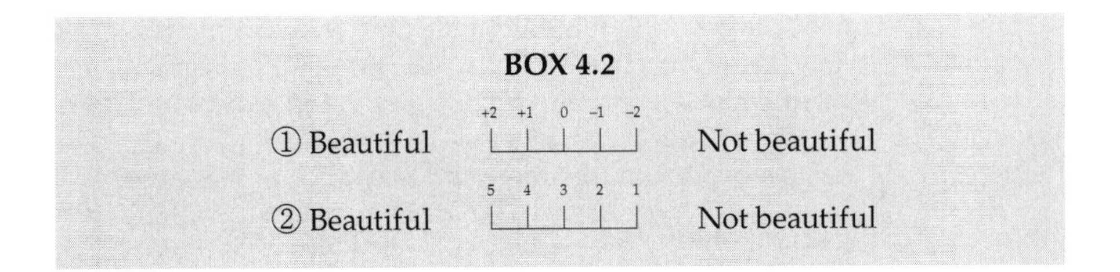

Normally, people tend to think that if they want to study in detail it is better to have more levels of evaluation. However, more rating levels will actually cause the test subjects to get confused, making it hard for them to evaluate, and in some cases, causing them to give haphazard evaluations.

In America, there was research on surveys using the 3-level method and 11-level method under the same conditions. The analysis result showed that there was no difference between those methods. Therefore, there is no need to argue about how many levels should be used in a rating scale. As a conclusion, it is most desirable to have a rating scale that is easy for the evaluators (test subjects) and based on common sense; the 5-level rating method has been used most widely.

There might be concern on how to score the evaluation. Two methods are as follows (see Box 4.2): (1) the No opinion position at the center is set as 0, the Beautiful side as plus, while the Not beautiful side as minus, and (2) the Beautiful side is the highest score and the scale gradually goes down toward the Not beautiful side. If we are going to plot a graph, (1) is convenient. For computer processing, (2) seems to be more convenient. There is no reason why we must use one and not the other.

4.4 Primary Evaluation Experiment

We already have about 600–800 pairs of Kansei words. Now we are going to start the experiment. Even if we could not get that many Kansei words, we still could perform the experiment, because this primary evaluation experiment is performed to select Kansei words for the actual experiment (secondary evaluation experiment), which will be explained later. Therefore, even with 300 or 400 Kansei words, this primary evaluation experiment still can be performed. However, it is important to collect as many Kansei words as possible.

Regarding the specimen for evaluation, you can choose a product from the genre that you or your company plan to develop as a new product. Here, as an example, let's choose a pot.

In this experiment, we have to collect all kinds of pots, including products from rival companies. However, since the purpose is to grasp the characteristics of Kansei words and select the words for the actual experiment, it is sufficient to collect about 10–20 types.

There are three ways to perform this primary evaluation experiment:

1. Arrange the actual products and perform the evaluation experiment while touching them.
2. Perform the evaluation experiment while showing a slideshow.
3. For large-sized products, prepare miniatures and evaluate by inspecting them.

Out of these, the best method is performing the evaluation while looking at and touching the actual products. Should there be no special constraint, I recommend an evaluation experiment using the actual products. What I mean by a special constraint is, for instance, the difficulty in arranging the objects such as houses, buildings, bridges, and towns. Even for these kinds of objects, we still can perform the evaluation by going to see the actual objects. However, there might be some difficult cases if we have a large number of test subjects (evaluators). In such cases, it is possible to use a slideshow as a substitute for the actual object, using photos taken with care so that the element to be evaluated can be accurately viewed from the slide. This also has something to do with the output of the slide. We made a comparison between the slideshow using photos taken with care and the actual objects for 20 types of rooms. The correlation coefficients of SD evaluation for both methods are between 0.880–0.932; therefore, even with a well-prepared slideshow, we still can effectively perform the evaluation experiment.

However, regarding electric pots, as well as writing instruments like a pencil, which will also be described as an example, they are products we need to touch in order to use. We cannot assume that carrying out evaluation experiments using slides for these items will be appropriate. Due to difficulties in bringing the actual objects together, plus the incurred cost, some suggest using miniatures for the evaluation experiment, but the size of miniatures can be a big problem. No matter how exquisitely a palm-sized miniature is made, it is not suitable for an evaluation experiment. If we were to make a sufficiently large and exquisite miniature, it would be very costly. In conclusion, I think it is better not to use miniatures. Furthermore, from the standpoint of evaluators, the psychological mood of looking at miniatures might distort the evaluation.

We arranged 10 units of electric pots and evaluated them using 600 Kansei words. The participants for the experiments were 40 people, consisting of R&D staff and designers. The age range of the participants was 23–40 years old, and the numbers of male and female participants were the same. An experiment that has such conditions is an excellent experiment.

Details on how to perform the experiment will be explained here. It is desirable to break down the whole process into a few sessions that will not exhaust the evaluators and to provide breaks between sessions. In general, a 15-minute break for every 2 hours of sessions seems to be appropriate. For a rough guide, if Evaluation subject × No. of Kansei words = 1000, then the session will take 1 hour. The calculation involved here is, if 10 units of subjects × 600 words = 6000, then the experiment will take approximately 6 hours; 20 units of subjects × 100 words = 2000, will take roughly 2 hours.

4.5 Factor Analysis of Kansei Words

I mentioned earlier that the objective of the primary evaluation experiment is to select Kansei words for the actual experiment. Another objective is to grasp the semantic space related to the subject field. For this purpose, we analyze the result of the primary evaluation experiment by factor analysis.

Factor analysis is one type of multivariate analysis. It is a statistical technique used in nonstructured data analysis to turn the multidimensional factors into a structure in a lesser dimension.

Let's say we got the factor analysis result for the 600 words related to the aforementioned electric pots as shown in Table 4.1. In Table 4.1, six factors were obtained, and the accumulated contribution was 0.850. If we obtain a high value of the accumulated contribution with lesser factors, we can say that it is a well-identified evaluation experiment. In the table, Kansei words 1–7 loading factors become high with Factor 1, and Kansei words 8–10 become high with Factor 2. Most statistics software will sort the high loading values for each factor like this, so they will create a list similar to the sample data of Table 4.1.

Next, we are going to give each factor a name that contains a common meaning that covers all the Kansei words with high loading values. Figure 4.4 shows the terms for factor analysis of a suit from the point of view of female university students. In this instance, nine factors are arranged in the sequence of the magnitude of contribution, starting from *high grade factor* until *advancement factor*. These nine factors denote the important pillars that make up the design of the clothes. In other words, it suggests that a suit lacking these nine factors means a suit design of poor Kansei, especially the top contributing factors like *high grade, high sense, active, luxuriousness, feminine,* and so forth, which are definitely essential characteristics.

Factor analysis helps us clarify the Kansei semantic structure of a subject like that in Figure 4.4, and it is a convenient way to gather suggestions about concepts in creating new products.

Osgood picked up three factors from factor analysis results and named the first factor the *evaluation factor,* the second the *active factor,* and the third

TABLE 4.1

Sample Factor Analysis for Electric Pots

Kansei Word	Factor 1	Factor 2		Factor 5	h²
(1)	.875		....		.785
(2)	.862				.921
(3)	.836				.888
(4)	.817				.767
(5)	.795				.809
(6)	.631				.905
(7)	.586				.803
(8)	⋮	.888			.792
(9)		.852			.778
(10)		.769			.826
⋮					
(597)					
(598)					
(599)				.685	.591
(600)				.663	.687
Contribution	.350	.210		.123	
Acc. Contribution	.350	.560		.887	

the *dynamism factor.* These names are based on the common meaning of the variables (in this case, words) that have high loading values for each factor. Subsequently, most researchers followed Osgood's naming for these three factors, but this is wrong. Osgood himself and many other researchers had extracted factors with different names. If the evaluation subject or the language used is different, the factor analysis result too will be different. Of course, the naming also will be different. Based on the results of each, analyze the Kansei words that belong to the axis of each factor. It is important to focus on giving names that represent those words.

Similar to the electric pot experiment, we can have the same idea as in Figure 4.4. If we give names to the six factors in Table 4.1, the Kansei semantic space for electric pot will become clear, and the structure that makes up a product called an electric pot will be made clear. Even without practicing Kansei engineering, by looking at the factorial structure and creating a new design (including functions) without omitting those factors, especially the top ones, a new product can be produced satisfactorily.

Kansei semantic space using factor analysis as described here is naturally constructed in the head of experienced designers who have been involved in new product development. Designers can tell just by gut instinct that certain things are absolutely necessary concepts in an electric pot design. They are working while thinking about things like how to

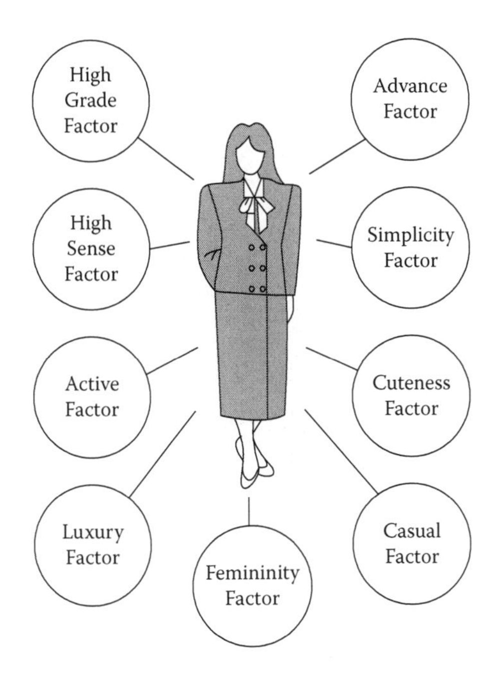

FIGURE 4.4
The nine factors of clothes design.

express the indispensable factors for new products, and then how to add value-added quality into the new design. When these concepts of design are expressed scientifically and numerically, we can derive the Kansei semantic space from factor analysis (Figure 4.5).

4.6 Extracting High-Level Kansei Words

In this step, we are going to select Kansei words for the actual experiment. In our experiment, 600 Kansei words have already been prepared, and based on these, we will select only the appropriate Kansei words. The reasons for this are as follows:

1. It seems unlikely that a consumer has 600 kinds of Kansei related to the product.
2. When performing the actual experiment, we will have to perform evaluation experiments on more types of specimens. Therefore, we want to narrow down the Kansei words to only the most necessary ones (Figure 4.6).

FIGURE 4.5
Factor analysis of Kansei words.

FIGURE 4.6
Item/category classification.

That is the key point in performing the primary evaluation experiment. We have already understood the Kansei semantic structure of the product that we are targeting through the primary evaluation experiment. It is important that the Kansei words cover the structure. If the Kansei words are extracted from all the factors, then the reproduction of the Kansei semantic space will be possible.

Kansei words are selected in the following ways:

1. Select a few from the Kansei words that are related to each factor. If possible, narrow down to only the words that have a similar meaning, and select the words with a different dimension in meaning. However, since the words are for the same factor, they will be similar. That is where the correlation coefficients sometimes come in handy. For words in the same factor axis, try to select the one with the lower correlation coefficient. Numbers of Kansei words in each factor axis are different. For example, the first factor naturally has a larger number of words, and the number becomes less and less after each factor axis. Therefore, we select proportionately the numbers of words that belong to each factor axis. Select more words from the factors that have many words, and fewer words from the factors with fewer words.
2. Having said the above, the most important selection method is selecting the desired words to best express the Kansei for the targeted product group. For example, if the words that consumers frequently use when purchasing the product or the words that the product designers want to use are not in the list, then it is meaningless. One way to overcome this is by showing the aforementioned 600 Kansei words and the factor axis to the R&D personnel and designers, and asking them to select the words that in their opinion are necessary.

The best way is to select the Kansei words using the combination of Steps (1) and (2).

The next issue is, what is the appropriate numbers of words? It is best to use the whole 600 words, but if we think of the logic of execution of the secondary experiment, that will be difficult. So, for the experiment to be reasonable, the number of words should be around 50–100 at most. Let's say we have chosen 100 words. It is important to confirm that all the words required by designers as in Step (2) are included. If they are, then we have a complete set of 100 Kansei words. Table 4.2 shows the 42 Kansei words obtained for car exterior design.

4.7 Configuration of SD Scale (Part 2)

Arrange the final selected 100 Kansei words in Osgood's SD scale format. As mentioned earlier, set the format to positive—negative, like *elegant—inelegant*,

TABLE 4.2

Sample Evaluation Sheet with 42 Kansei Words

		No.__________	Name__________________
1	Cute	□□□□□□	Not Cute
2	Textured	□□□□□□	Not Textured
3	Variable	□□□□□□	Invariable
4	For Street Racers	□□□□□□	For All Users
5	Dynamic	□□□□□□	Static
6	Maniac	□□□□□□	Not Maniac
7	Futuristic	□□□□□□	Retro
8	Unwearying	□□□□□□	Wearying
9	Charming	□□□□□□	Charmless
10	Flat	□□□□□□	Not Flat
11	Fancy	□□□□□□	Not Fancy
12	Calm	□□□□□□	Not Calm
13	Unstrained	□□□□□□	Strained
14	Emotional	□□□□□□	Not Emotional
15	Intellectual	□□□□□□	Wild
16	Individual	□□□□□□	General
17	Masculine	□□□□□□	Feminine
18	Refined	□□□□□□	Unrefined
19	Good Design	□□□□□□	Poor Design
20	Well-organized	□□□□□□	Chaotic
21	Unity	□□□□□□	No Unity
22	Luxury Car	□□□□□□	General Car
23	Elegant	□□□□□□	Inelegant
24	Casual	□□□□□□	Formal
25	Adult-like	□□□□□□	Young-like
26	Sophisticated	□□□□□□	Unsophisticated
27	Non-persisting	□□□□□□	Persisting
28	Ponderous	□□□□□□	Thoughtless
29	Have Playful Mind	□□□□□□	No Playful Mind
30	Humorous	□□□□□□	Insipid
31	Innovative	□□□□□□	Conservative
32	Western-style	□□□□□□	Japanese-style
33	Beautiful	□□□□□□	Not Beautiful
34	Favorable	□□□□□□	Unfavorable
35	Sporty	□□□□□□	Not Sporty
36	Feel the Wind	□□□□□□	Not Feel the Wind
37	Youthful	□□□□□□	Not Youthful
38	Rounded	□□□□□□	Rectilinear
39	Good Looking	□□□□□□	Bad Looking
40	Have Volume	□□□□□□	No Volume
41	Want to Ride	□□□□□□	Don't Want to Ride
42	Fast	□□□□□□	Slow

BOX 4.3

	Very			Neither			Very	
Elegant	☐	☐	☐	☐	☐	☐	☐	Inelegant

not elegant—crude. If the opposite word does not carry a negative value, such as *bright—dark*, the pair can be used as it is.

The evaluation scale can be either a 5-level rating or 7-level rating. For the case study, I will use the 7-level rating example (see Box 4.3).

In this example, for Neither, draw a circle in the center box of the seven levels. Mark the leftmost box for Very elegant, the next box for lesser degree of elegant, and the box on the left side of the Neither level to Slightly elegant.

It is recommended to rehearse using just one word before starting the evaluation experiment with the 100 Kansei words. Select one word that is not in the 100 words as an example, and perform this preliminary simulation prior to the experiment.

4.8 Secondary Evaluation Experiment

This is the most important step in the creation of the database. Since the evaluation subject is an electric pot, collect various designs of pots by all makers. Let's say 20 types of pots are collected. This means we are going to evaluate those 20 types of pots using the aforementioned 100 Kansei words.

First, regarding the Kansei words evaluation form, if the 100 words are arranged in the same sequence, toward the end of the 20 pots the evaluators get tired and bored, which might cause them to make a haphazard evaluation. In order to eliminate this, use the Kansei words evaluation forms arranged in different sequences for each group of five test subjects. The most desirable method is all the test subjects evaluate the randomly arranged objects using the randomly printed Kansei words, but the preparation as well as the analysis for that will be very difficult. The possibility for error will also be high.

A simple way to eliminate the effect of order is to create groups of 10 Kansei words, change the order of the groups, and distribute the same survey form to a group of five people. By doing this, the effect of order will disappear, and at the same time, it is also possible to survey about 40–50 test subjects at one time.

Put labels A, B, C … on each electric pot, and give evaluators the freedom to touch and see them while evaluating, using the 100 SD scale words to convey one's own feeling (Figure 4.7).

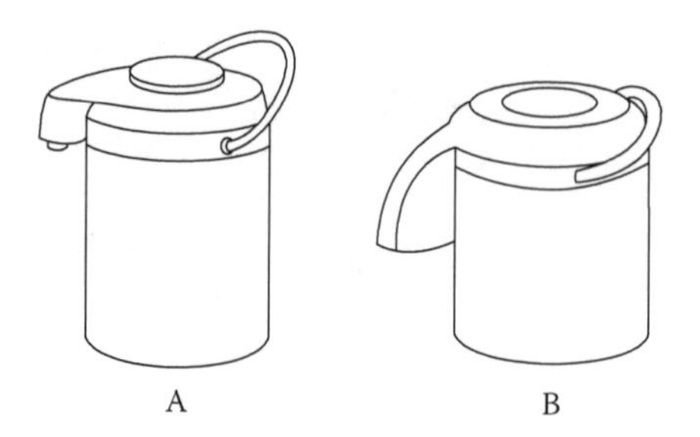

FIGURE 4.7
Two types of electric pots as an example.

4.9 Preparation of Evaluation Samples

Regarding how to select the evaluation samples for the primary evaluation experiment, there's no need to worry. The purpose of the experiment is to narrow down the 600 Kansei words to 100 (or maybe fewer) to perform factor analysis.

However, for the secondary evaluation experiment, since we are going to use the results to create a database, it is necessary to include whatever samples are available. Some points of caution in collecting the samples are as follows:

1. If we want to survey the appearance design, collect as many samples as possible that have a different appearance. One type is enough for samples of similar shape.
2. If we want to survey mainly the color, collect as many samples as possible that have a similar appearance with different colors (color, floral motif, etc.).
3. If we want to survey the appearance as well as the color, we need to collect as many different types of samples that contain both attributes. In this case, we eventually have to collect quite a number of products.

Since we utilize existing products and perform the Kansei survey, it will gradually become clear what kind of Kansei are associated with the characteristics contained in different products.

Since we only know the characteristics of the existing products here, there is a concern whether we can really know the predictive Kansei of a future product. However, as I will explain later, by using the quantification theory, which is one of the multivariate analyses in statistical processing, we can know every characteristic and the Kansei of those combinations. Therefore, by considering the combinations from the designer's viewpoint, future predictions could be made.

However, if you wish to study the future design more thoroughly, one possible solution is to use freehand drawings to replace the actual product as the evaluation subject. By preparing and adding the future design drawings to the collection, it is possible to create a future prediction database. In fact, at Hiroshima University, such a research trend has begun with the cooperation of industrial designers.

4.10 Extracting Item/Category

We have completed the Kansei words SD scale of the kind of words to describe the Kansei. So, the next step is the design analysis of the product specimen. We will go into a process called extraction of design item/category.

Design *item* refers to a certain characteristic in product design. For example, in the case of an electric pot, height, width (diameter), color, and so on are the design items. *Category* refers to the small groupings in each item, which is the internal classification such as (1) 25–28 cm, (2) 28–31 cm, (3) 31–34 cm, (4) 34–37 cm, and (5) >37 cm for the height item. Regarding the pot's height, collect a number of pots, measure their height, subtract the minimum value from the maximum, divide that into 3–4 segments, and set the category. We can set the category for the diameter and so forth in the same way.

However, some items cannot be expressed in a physical value such as height, for example, *small curve* or *big curve*, or *not many functions* or *many functions*. Length and height are called quantitative characteristics, while something that cannot be described numerically like *big/small floral motif* are called qualitative characteristics. Various aspects of qualitative characteristics are involved in the design of a product, and since these Kansei are also subtle influences, it is necessary to include their analysis as well.

For the electric pots, a segment from the actual example of item/category classification is shown in Table 4.3. The first 1–4 are quantitative characteristics, and the rest are qualitative. In general, if we can categorize more than 100 categories, we will get a good result. If we get about 20–30 categories, some important design elements might be left out, and even when we perform statistical processing, there is a possibility that it will yield a meaningless output.

The secret to classifying the item/category is to keep the perspective of the user's viewpoint on the design. If we make the classification only from the designer's viewpoint, it will turn out to be merely unit division. On the other hand, users will break down certain elements and take a global view for certain other elements. Classifying the design from the user's viewpoint is the essence of a Kansei study.

For example, users do not see the height or width of the electric pots technically. They will combine the two elements and see them as *short and stout*

TABLE 4.3

Example of Item/Category for Electric Pots

Item	Category	Item	Category
1. Height	① 25–28 cm ② 28–31 cm ③ 31–34 cm ④ 34–37 cm ⑤ >37 cm	9. Lid design	① Integrated type ② Separated type
		10. Outer shape design	① Short and stout type ② Medium type ③ Thin and tall type
2. Width	① 14–16 cm ② 16–18 cm ③ >18 cm	11. Water level	① Clearly visible ② Hardly visible
3. Height/width	① 1.5–1.75 ② 1.75–2.12 ③ >2.12	12. Lid curve	① Small ② Medium ③ Big
4. Projecting rate of spout design	① 0.5–0.7 ② >0.7	13. Cord	① Retractable ② Not retractable
5. Body shape	① Round ② Oval	14. Appearance color	① White ② Light pink ③ Red ④ Black ⑤ Others
6. Spout shape	① Big elephant shape ② Small elephant shape ③ Cap shape	15. Pattern	① No ② Geometric pattern ③ Floral motif ④ Others
7. Pouring function	① Air type ② Electric type		
8. Functions	① Few functions ② Many functions	16. Floral motif	① Not floral motif ② Small floral motif ③ Medium floral motif ④ Big floral motif

or *thin and tall*. Another example is car interior. Whether a human will feel the *spacious feeling* or not depends on the consistency of the flow of left door + instrument panel + right door. If the design does not create a sense of an integrated flow between the doors along the instrument panel—if the instrument panel is disconnected from the doors—it will not create a spacious feeling.

Let's say we have collected 20 electric pots. We must prepare the item/category table for the pots. Mark a circle in the appropriate boxes according to the aforementioned Table 4.3 for the pots in the sequence of A, B, C…, to make a list as shown in Table 4.4.

We have to be careful here that

1. There should be only one circle marked in an item for each pot.
2. There should be no products with the same circle marks for all items/categories.
3. At least two units must have a circle mark in each category.

If 1–3 are not strictly complied to, the computer will stop the statistical calculation. This detail will be explained later.

TABLE 4.4

Item/Category Table

Item	Category	A (a Co.)	B (a Co.)	C (p Co.)	D (q Co.)	E (b Co.)	F (e Co.)	G (f Co.)	H (f Co.)	
1. Height	① 25–28	○		○						2
	② 28–31		○				○			2
	③ 31–34				○					1
	④ 34–37							○	○	2
	⑤ >37					○				1
2. Width	① 14–16			○				○		2
	② 16–18	○			○		○		○	4
	③ >18		○			○				2
3. Height/width	① 1.5–1.75	○		○				○		3
	② 1.75–2.12				○		○			2
	③ >2.12		○			○			○	3
4. Projecting rate of spout design	① 0.5–0.7	○		○			○			3
	② >0.7		○		○	○		○	○	5
5. Body shape	① Round	○	○	○	○		○	○		6
	② Oval					○			○	2

4.11 Statistical Analysis

4.11.1 Factor Analysis of Kansei Words

We have performed factor analysis for 600 Kansei words. Here, we are going to perform another factor analysis, but for the reduced number of Kansei words. There are various types of factor analyses. You can pick any of them.

When we perform factor analysis, the factor axes that are almost the same as the 600 words should appear. This is because we had selected a few words from each factor axis of the 600 words. Normally, a factor analysis is configured to stop its operation when the eigenvalue becomes lower than 1. Check the accumulated contribution. A high value like 0.8 means reliable, but if it is around 0.5 or 0.6, this shows that there are still significant factor axes being left out. The reason might be that the correct Kansei words to extract the Kansei were not included, or that there were special people among the test subjects.

4.11.2 Analysis of SD Evaluation

We have evaluated the 20 electric pots using the SD-scaled Kansei words. Now, let's check what kind of image has been discovered for each pot using an SD diagram. Figure 4.8 shows the SD diagram for two types of pots as in Figure 4.7, evaluated using 25 Kansei words as an example. Since these 25 Kansei words are for pots, I think they can be used as a reference.

As we can see, pot A is a thin and tall type, while pot B is short and stout type. The average value of SD scale for a few dozen test subjects is shown in Figure 4.8. If we look at the characteristics of pot A, the scores are high for many words, such as *cute, unweary, refined, sophisticated, humorous,* and *neat.* On the other hand, pot B does not have many characteristics, and its images are *flat, uneven, poor design, inelegant, not classy, mature,* and *outdated.*

How this kind of image difference appears depends on the mechanism of the item/category analysis of the product. Pot A, which is a thin and tall type with a neat design, looks *simple, unwary,* and *neat.* Therefore, it portrays a *refined* image. On the other hand, pot B, which is short and stout with an elephant-like big spout, looks *inelegant* and *not classy.* Since it has been used for a long time, it appears to be *outdated.* Such effort to interpret the differences is needed when performing the Kansei engineering analysis. It is an method that relates the Kansei differences to the derived design differences.

For the 20 electric pots, we prepared an SD diagram as in Figure 4.8 to check for any exceptional expressions by looking at the distribution of average value of test subjects. Additionally, check all the original evaluation records by test subjects for abnormalities such as those who put circles only in the middle (no opinion), those who have a totally different response from

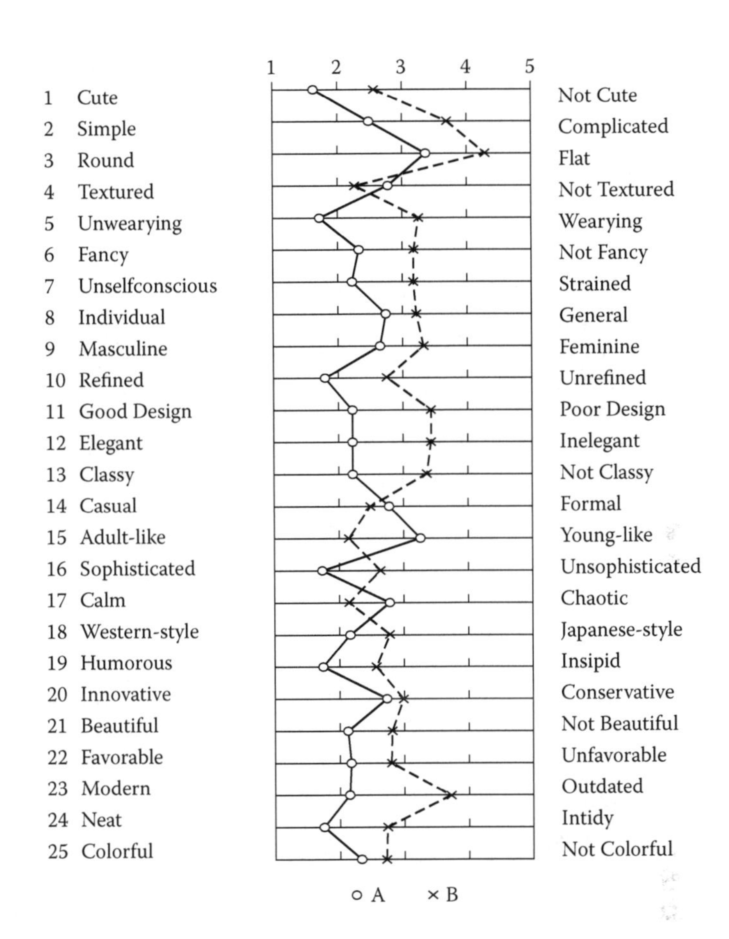

FIGURE 4.8
Evaluation results of the pots using 25 Kansei words.

others, and other abnormalities. If any fall into these categories, delete the data and regard it as abnormal or invalid.

4.11.3 Multivariate Analysis

The evaluation values obtained from the SD scale are calculated using multivariate analysis to prepare the foundation for the database. Now, I will give some explanations about multivariate analysis. The above-mentioned factor analysis is one type of multivariate analysis. If something is decided for one reason only, it is called univariate. If a phenomenon occurs, or various characteristic values are obtained for multiple reasons (variables), it is called multivariate. Multivariate analysis is an technique used to clarify the impact of variables that affect the characteristic values as a phenomenon. Factor analysis is also a method using variables, that is, the factor axes.

Multivariate analysis can be divided into (1) predicting characteristic values (objective variables or external criteria) based on information that has few variables (explanatory variables), and (2) structuring the data that do not have an objective variable (external criterion) using explanatory variables. Then, this can be further divided into (3) one in which the data are given in numerical values, and (4) one with qualitative data such as *like* or *don't like*. Which technique of multivariate analysis to use for (1) or (2), or for data such as in (3) or (4), can be diagrammed as in Figure 4.9.

Explaining everything in Figure 4.9 might cause confusion, so I will explain only the techniques that are most used in Kansei engineering. First, as mentioned before, factor analysis refers to an analytic method that structures the data that do not have objective variables, making it easier to read out the data structure. For example, in order to know the Kansei data structure for the clothes of female university students, the clothes are evaluated using Kansei words, and the result of the factor analysis is as shown in Figure 4.4. Here we had nine factor axes; this means we have discovered that the Kansei of clothes for young female university students is a Kansei constructed of nine pillars. All the Kansei (called explanatory variables) can be explained using these nine pillars (factors). This has been called *data structure* and also *semantic space*.

The semantic space of the clothes consists of nine factor axes. This means that when making clothes for female university students, it is important to include these nine factors into the clothing composition. Additionally, if we want to produce *intellectual* clothes, since we know the degree of impact of the *intellectual* Kansei among these nine factors, we need to consider the factors with a high degree of impact. Otherwise we will not get *intellectual* clothes. In this sense, the factor axes of a factor analysis provide us with important information.

Next, the quantification theory Type I is used in Kansei engineering. This is an analytic method in which a phenomenon determined by the external criteria using the classification format is predicted using the quantitative explanatory variables. For example, in order to clarify what kind of explanatory variables (in this case, the design elements) are effective in elicitation of the *elegant* and *inelegant* electric pots, we perform analysis using the quantification theory Type I.

I will explain how to utilize the quantification theory Type I in Kansei engineering using the example of a pencil as a writing instrument.

4.11.4 How to Utilize Quantification Theory Type I

We selected seven types of pencils with different designs and evaluated them using 15 Kansei words as shown in Table 4.5. The table shows the Kansei words collected by designers at pencil manufacturers. The participants in the evaluation experiment were the designers.

Prior to the evaluation using the SD scale, classification for the item/category of the pencil designs was performed. The design characteristics of

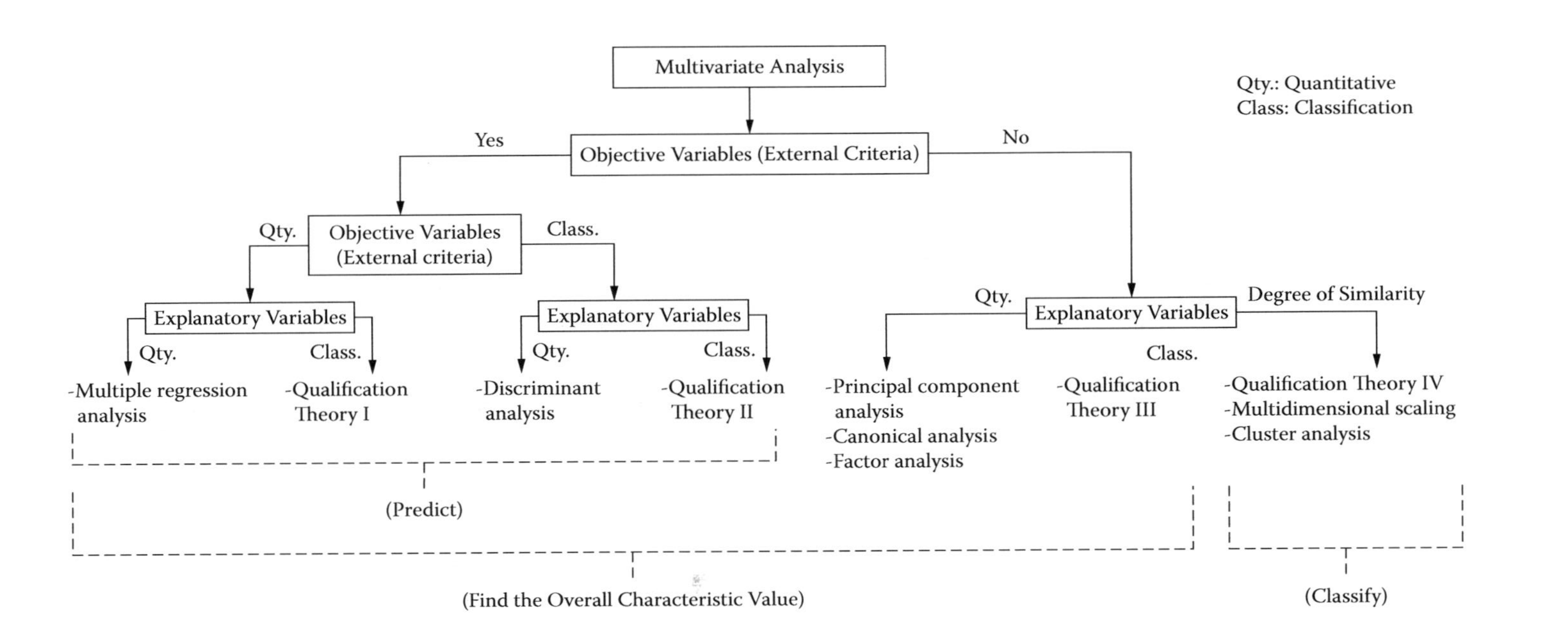

FIGURE 4.9
Classification of multivariate analysis.

TABLE 4.5
Evaluation of Pencils Using Kansei Words

1.	Not Colorful	□□□□□	Colorful
2.	Not Smart	□□□□□	Smart
3.	Not Authentic	□□□□□	Authentic
4.	Not Mechanical	□□□□□	Mechanical
5.	Cold	□□□□□	Warm
6.	Cheapie	□□□□□	Classy
7.	Boorish	□□□□□	Fashionable
8.	Rough	□□□□□	Precise
9.	Hard	□□□□□	Soft
10.	Not Fit in Hand	□□□□□	Fit in Hand
11.	Gloomy	□□□□□	Bright
12.	Decorative	□□□□□	Simple
13.	Old	□□□□□	New
14.	Rounded	□□□□□	Rectilinear
15.	Traditional	□□□□□	Advanced

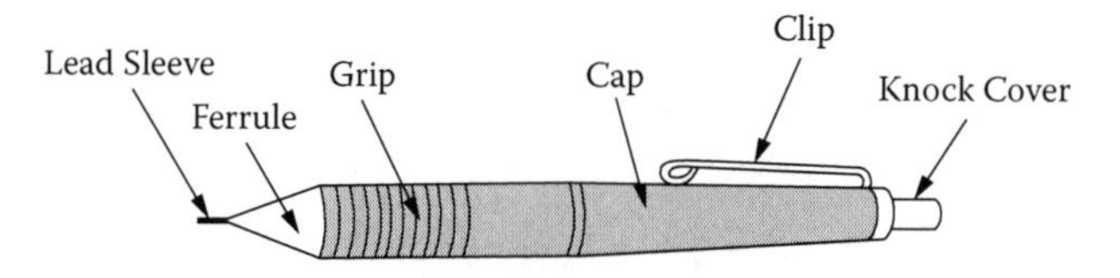

FIGURE 4.10
Quantification theory Type I for pencil: *Not authentic—Authentic.*

a pencil are named in Figure 4.10, and the classified item/category as well as the average value for each classification of pencil evaluated using each Kansei word are shown in Tables 4.6 and 4.7, respectively.

Let me explain the analysis result when the quantification theory Type I is performed for these data. First of all, the result of external appearance criteria for an *authentic pencil* is shown in Table 4.8. To read the table, first of all, check the level of the multiple correlation coefficients. If it is above 0.8, it shows that it can be sufficiently explained using the explanatory variables considered.

Next, check the partial correlation coefficient. This partial correlation coefficient is equivalent to the degree of contribution for classifying *authentic* and *not authentic*. From Table 4.8, the authentic pencil is described as having:

1. Two colors
2. Achromatic basic color tone (white, black)
3. Cylindrical-shaped body without step
4. Independent type clip
5. Metal cap of standard diameter

TABLE 4.6

Item/Category Classification of Pencils

Item	Category/Sample	A	B	C	D	E	F	G
1. Lead sleeve and ferrule	① Exposed type		○	○		○	○	
	② Retractable type	○			○			○
2. Grip and cap	① Accordion shape, no step	○						
	② Cylinder shape, no step		○			○	○	
	③ Cylinder shape, has step			○	○			○
3. Clip	① Integrated type			○	○		○	○
	② Independent type	○	○			○		
4. Cap	① Metal coated, standard type		○			○		
	② Nonmetal coated, standard type	○		○	○			
	③ Resin, extrathick						○	
	④ Resin, transparent, standard type							○
5. Color	① One color		○	○			○	○
	② Two colors	○			○	○		
6. Color base tone	① Red tone	○		○		○		
	② Blue tone							○
	③ No color		○				○	

TABLE 4.7

Kansei Average Value for Each Pencil

Kansei Word/Sample	A	B	C	D	E	F	G
1. Not colorful—Colorful	2.6	1.0	4.4	3.0	4.4	1.0	4.0
2. Not smart—Smart	3.4	4.4	3.2	3.2	3.0	1.2	4.0
3. Not authentic—Authentic	3.0	3.2	1.6	2.6	3.8	2.0	1.4
4. Not mechanical—Mechanical	4.6	4.0	2.0	2.6	4.4	1.8	3.4
5. Cold—Warm	3.2	1.4	4.2	3.4	3.0	3.4	3.2
6. Cheapie—Classy	3.8	3.2	1.8	2.4	3.8	2.4	1.4
7. Boorish—Fashionable	3.6	2.6	3.2	3.4	3.4	2.6	2.8
8. Rough—Precise	3.8	4.0	2.4	2.6	4.0	2.6	1.6
9. Hard—Soft	2.8	1.2	4.0	3.6	2.0	4.4	3.4
10. Does not fit in hand—Fits in hand	4.2	2.2	3.4	3.6	2.8	4.6	1.6
11. Gloomy—Bright	2.2	2.2	4.2	3.8	3.6	2.0	4.2
12. Decorative—Simple	2.6	4.6	3.8	4.0	3.0	4.0	4.2
13. Old—New	3.4	2.0	3.4	3.4	3.2	3.4	3.2
14. Rounded—Rectilinear	3.4	5.0	2.4	3.2	4.4	2.2	4.4
15. Traditional—Advanced	3.4	2.2	3.4	3.2	4.2	3.2	3.4

TABLE 4.8

Item/Category in FAIMS

Item	Category	Item	Category
Coat		**Blouse**	
Collar	No (Round shape) No (V-shape) Have	Collar	No Necktie V-neck Round neck Stand Ribbon Bow tie
Type	Single Double		
Length	Waist Half-length Hip	Hue	Red Orange Yellow Green Green rust Blue Bluish-purple Purple Reddish-purple Neutral
Clearance	No Have		
Decoration	No Button, line Belt		
Skirt			
Shape	Tight Flared Gathered Pleated	Tone	White Light gray Pale Light Bright Gray Light grayish Vivid Dark gray Dark Deep Black
Length	Knee-length Normal length Long		
Decoration	Button Box Other No		
Fabric pattern	Have No		

In other words, it is made of metal, it is white, of about 0.8 mm cylindrical shape, and neat in design, which refers to the sample of type B. This is the design that is then produced in all countries as a high-class pencil.

Next, let's have a look at a *fancy pencil* (Figure 4.11). This pencil is designed with:

1. Red color tone
2. Two colors
3. Resin-coated of standard diameter
4. Independent-type clip
5. Accordion shape without step

Item	Category	Partial Correlation Coefficient	Score
1. Lead Sleeve and Ferrule	① Exposed Type	0.274	−0.171
	② Retractable Type		0.227
2. Grip and Cap	① Accordion Shape, No Step	0.686	0.257
	② Cylinder Shape, No Step		0.657
	③ Cylinder Shape, has Step		−0.742
3. Clip	① Integrated Type	0.714	−0.385
	② Independent Type		0.514
4. Cap	① Metal Coated, Standard Type	0.653	0.471
	② Non-metal Coated, Standard Type		−0.028
	③ Resin, Extra-thick		−0.128
	④ Resin, Transparent, Standard Type		−0.728
5. Color	① One-color	0.862	−0.661
	② Two-color		0.881
6. Color Base Tone	① Red Tone	0.759	−0.081
	② Blue Tone		−0.624
	③ No Color		0.746

FIGURE 4.11
Components of a *fancy* pencil.

In this case, the color provides an important effect. One of the colors is red, and the pencil has an accordion-shaped surface in the holding area to make it easier to hold.

Lastly, let's look at the result for a *pencil that fits (comfortably) in hand* (Figure 4.12 and Figure 4.13). The design elements that contribute to the impression are:

1. Extra thick resin cap
2. Independent type clip
3. Cylindrical-shaped body without step

So, it is evident that there are two designs that made the pencils fit in the hand. One is the extra thick pencil body with a diameter about 12–13 mm, and another one is a rubber cover that makes the gripping area thick. The achromatic color—white, gray, or black—seems to be a good option (Figure 4.14).

I have explained how to read the calculation result using the quantification theory Type I from the aspect of Kansei for pencils. Apart from this method, there are other statistical analyses used in Kansei engineering. However, the quantification theory Type I is a very convenient method, because:

1. The point that splits the external criteria, *authentic pencil* and *not authentic pencil*, will show the contributing design elements in

Item	Category	Partial Correlation Coefficient	Score
1. Lead Sleeve and Ferrule	① Exposed Type	0.113	0.042
	② Retractable Type		−0.057
2. Grip and Cap	① Accordion Shape, No Step	0.540	0.571
	② Cylinder Shape, No Step		−0.261
	③ Cylinder Shape, has Step		0.071
3. Clip	① Integrated Type	0.553	−0.128
	② Independent Type		0.171
4. Cap	① Metal Coated, Standard Type	0.804	−0.257
	② Non-metal Coated, Standard Type		0.342
	③ Resin, Extra-thick		−0.357
	④ Resin, Transparent, Standard Type		−0.157
5. Color	① One-color	0.909	−0.183
	② Two-color		0.244
6. Color Base Tone	① Red Tone	0.910	0.212
	② Blue Tone		−0.016
	③ No Color		−0.302

FIGURE 4.12
Quantification theory Type I for pencil: *Boorish—Fashionable.*

numerical figures. The partial correlation coefficients, which were explained earlier, will show the degree of contribution toward the classification of external criteria. And the plus or minus score shows on which side they reside. Based on these, if we pick up the design elements (the item/category) that have a high partial correlation coefficient and a high plus value, we can construct an authentic pencil. Therefore, we can easily assemble a design as a product strategy.

2. By setting the product strategy to *authentic pencil* or *fancy pencil,* we can easily assemble the item and category that make up the design. It is also possible to consider various strategies.
3. Here, we have a list of what kind of impact the item/category has on the external criteria. It is convenient, and we can easily create the design database, which will be explained later.

On the other hand, there are also a few problems, such as:

1. Kansei was originally something ambiguous and not definite. If we perform statistical calculations, the quantification theory Type I regards each item/category as something independent and composed of the sum of linear variables. However, the Kansei itself has

Item	Category	Partial Correlation Coefficient	Score
1. Lead Sleeve and Ferrule	① Exposed Type	0.274	0.342
	② Retractable Type		−0.457
2. Grip and Cap	① Accordion Shape, No Step	0.686	1.457
	② Cylinder Shape, No Step		−0.342
	③ Cylinder Shape, has Step		−0.142
3. Clip	① Integrated Type	0.714	−0.300
	② Independent Type		0.400
4. Cap	① Metal Coated, Standard Type	0.653	−1.100
	② Non-metal Coated, Standard Type		0.600
	③ Resin, Extra-thick		1.700
	④ Resin, Transparent, Standard Type		−1.300
5. Color	① One-color	0.862	−0.391
	② Two-color		0.522
6. Color Base Tone	① Red Tone	0.759	0.049
	② Blue Tone		−0.665
	③ No Color		0.591

FIGURE 4.13
Quantification theory Type I for pencil: *Not fit in hand—Fit in hand.*

FIGURE 4.14
Pen Kansei engineering.

a nature of nonlinearity, so it is insufficient to perform this kind of statistical processing.

2. Therefore, it is also insufficient to express the design of a product by organizing the design parts in mosaic.

Since we have this kind of problem, at Hiroshima University, we have been trying to perform the processing while maintaining the nature of nonlinearity by using the fuzzy quantification theory Type I, neural network model and genetic algorithm. Since the contents are quite complicated, I will not try to explain them in this book. Even though the quantification theory Type I presents this problem, it still has much merit as the initial multivariate analysis, so we will continue to utilize it as we did before.

4.12 Creation of Databases

Since the objective is to build an expert system (a type of artificial intelligence) in a computer and to display the images that match each Kansei, we need to create a few databases.

4.12.1 Kansei Words Database

To process the criteria for the product that best fits certain Kansei words, we built a computer database for storage of Kansei words. For example, for the electric pot, there were 25 Kansei words. We built a mechanism where the most suitable pot to whichever word we had chosen will be selected. Therefore, if we choose the word *cute*, the pot that best fits this word will be displayed on the computer.

Among the information related to Kansei words are:

1. The Kansei words themselves
2. The factor structure (each factor axis and factor loading) calculated from the evaluation result of Kansei words
3. The correlation coefficients between Kansei words

This information is stored in a data storage file. Here we have storage of Kansei words, each factor axis and its loading, as well as the correlation coefficients between Kansei words.

4.12.2 Design Elements Database

To show the complete form of a product over a Braun tube (cathode ray tube), the whole product needs to be determined exclusively. To calculate

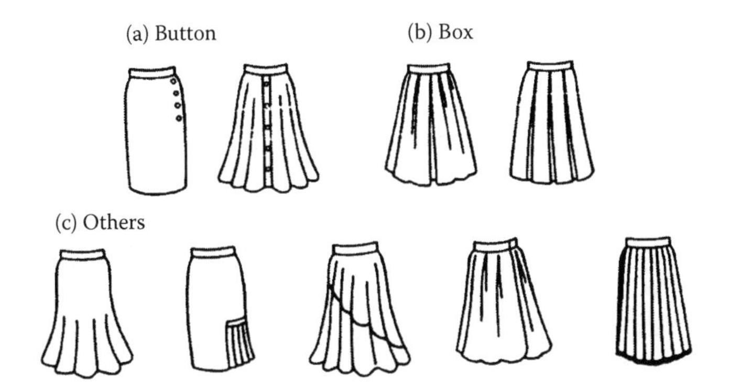

FIGURE 4.15
Parts for skirts shapes in FAIMS.

using the quantification theory Type I, we broke down the details of product design into item/category. However, we need to calculate and select in advance which category to assign, and demonstrate the product in the form of a graphic.

Thus, we created a system where, in order to demonstrate a product, the design elements are stored in a database in the form of item/category, and by a calculation, the corresponding product will be selected. A file name will be assigned to the item/category, and the file will be stored in the computer.

To make this easy to understand, I will explain using the fashion design for female university students (FAIMS), as shown in Figure 4.4. The item/category for the female university students' clothes are broken down as shown in Table 4.8. For example, the skirt shapes are (1) tight, (2) flared, (3) gathered, and (4) pleated. If we combine two types of skirts with buttons, two types of box pleats, and five types of others, we will have skirt shapes as in Figure 4.15.

In reality, more detailed combinations than these can be thought of. These are stored in the graphic database as one of the components. When a Kansei word is input, the appropriate components will be combined, and an illustration of a female university student, as shown in Figure 4.4, will be displayed on the computer screen.

Classifying the item/category, and creating various components by combining them, is the know-how in the success of Kansei engineering as an artificial intelligence system.

4.12.3 Rule Base

Rule base refers to a set of rules expressed using *if—then*. The inference engine of an expert system will perform inference using this rule base. In Kansei engineering, apart from the rules to resolve any conflict of Kansei word combinations and colors, most of the rules are from the calculation resulting from quantification theory Type I.

4.13 Artificial Intelligence (AI) Prototyping

In Kansei engineering, a customer will be asked to describe the images that he or she has in mind in the form of Kansei words. When those Kansei words are input into a computer, the image of the desired product will be displayed. This is the general format of Kansei engineering Type II. In order to create this mechanism, an expert system that is an application-type of AI is utilized.

The expert system refers to a system that embeds the knowledge of professionals (expertise) as its rule base and displays a conclusion after performing inference according to the rule. The professional knowledge is constructed in the *if—then* format. For most cases, it is written using the predicate logic, but also in some cases the numerical data file is used, like in Kansei engineering.

AI prototyping refers to the construction of a framework for an overall process using an expert system, and the data fed into it can be temporary or from a database. A blackboard or similar model is constructed as a method to check for conflicts and resolve them, if possible. When we know that all the components are working as expected, we will then go into the development of a powerful system.

4.14 Establishing an AI System

The establishing of the system is described as follows, from the beginning:

1. Adjectives database as the adjectives processor—A list of Kansei words related to a specific domain, as well as the loading factors and correlation coefficients of the factor analysis related to it is put into the system as a file.

   The input Kansei word is not always a single word. For FAIMS and HULIS, which were created by Hiroshima University, the customer can choose and input up to 10 words. For example, *luxurious and gorgeous* are on the same factor axis, so there will be no problem. However, for *luxurious but not flashy*, there is a slight conflict; so we need to create a mechanism to resolve the conflict. Methods such as the blackboard model or the fuzzy set can be used. We can also set the customer's psychological attributes to a rule and utilize them. Figure 4.16 shows the flowchart for the adjective processor.

2. Inference processor—The image database, which shows the association between Kansei words and design elements, and the rule base

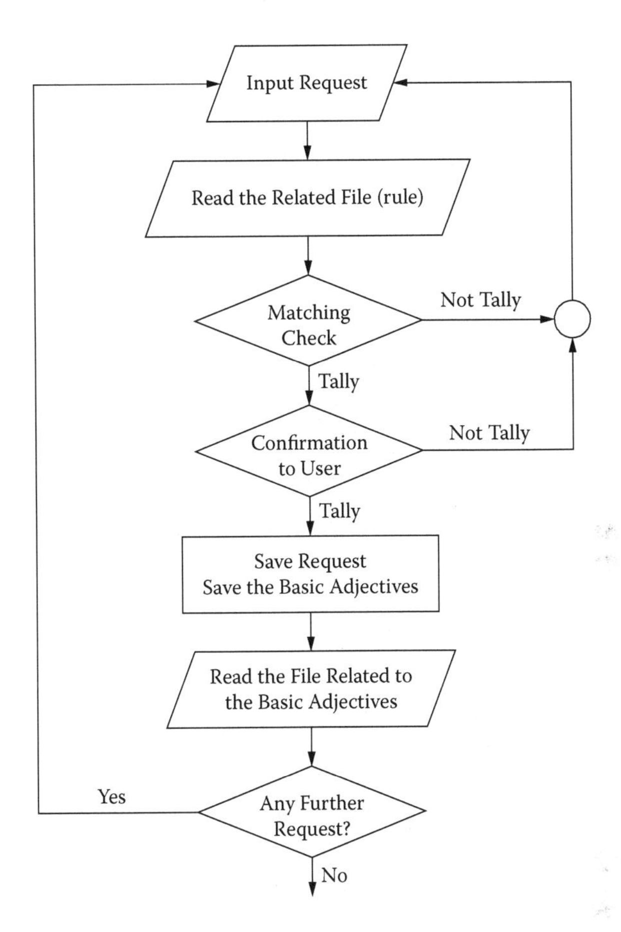

FIGURE 4.16
Flowchart for the adjectives processor.

(knowledge base) are built in this part. When a Kansei word is input, it will be verified, and the solution is obtained using the *if—then* condition. Colors especially tend to have conflicts, so this processor will perform conflict solving using a separate blackboard model or will select a correct conclusion using the rule based on the color condition.

3. Graphic display—Conclusions are made using codes. The design and color that are matched to the color code are confirmed from their respective databases, combined together, and displayed on the monitor. As mentioned earlier, the design elements (item/category) are stored in the shapes and colors database in the form of parts; the result of those combinations will be displayed as an image. The schematic drawing for the graphic display is shown in Figure 4.17. There is a function to change the details if they do not match the image.

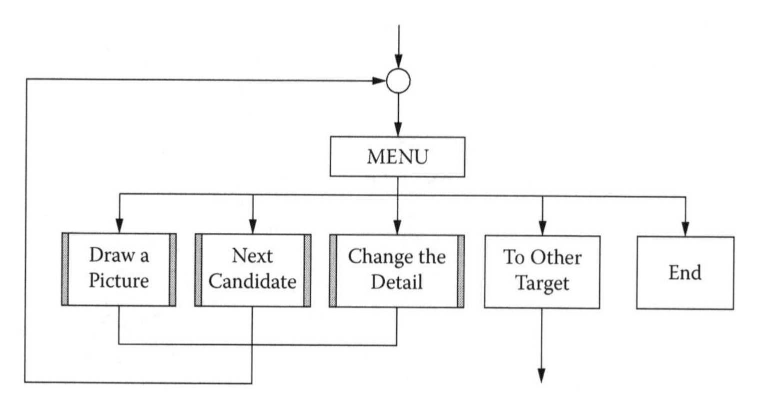

FIGURE 4.17
Schematic diagram of the graphic part.

4. System control—Make sure those three processors (1–3) are working properly, and then build up the control mechanism as a system.
5. Client interface—The Kansei engineering system will differ depending on who will use it, and we will prepare an interface to make it easy for the user (client) to understand and to use. We first need to predict what the users think and what they want to do, such as how to input, how to make inquiries, response while waiting, how to display images, how to make simple corrections and amendments, and how to proceed to the next step. Then we build a desirable interface. For example, if the users are designers, then facilitating product development is especially desired because it seems that designers in general do not really like computers.

Figure 4.18 shows the general structure of a Kansei engineering system made up with thorough conditions as mentioned here. It is a system that translates user Kansei into designs; therefore, it is called a *forward Kansei engineering system.* The contrary AI system, which diagnoses Kansei from designs, is called a *backward Kansei engineering system.* A system that includes both is called *hybrid Kansei engineering.*

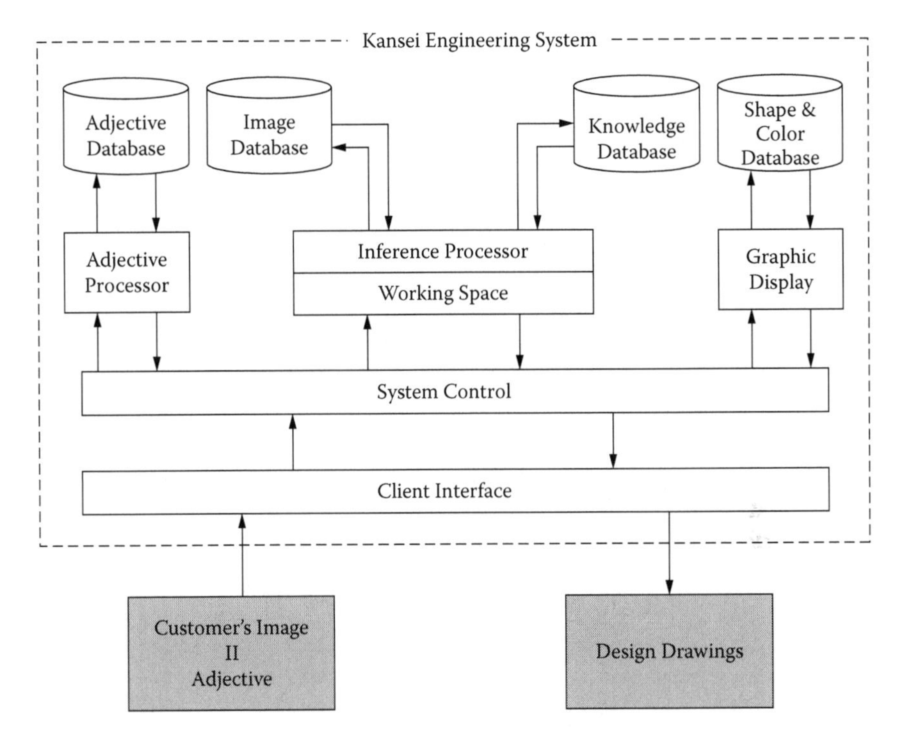

FIGURE 4.18
Kansei engineering system (forward).

5

Kansei Engineering Type II Application Cases

5.1 HULIS

HULIS, short for *human living system*, was the first Kansei engineering system. Until its introduction, I saw many conflicts between homeowners and builders. For example, the wallpaper that the owner had in mind, based on a small sample, was totally different from the actual wallpaper provided. The architect did not really make an effort to fulfill the owner's desire. I created a system that translates the consumer's (owner's) Kansei to a house design. That is HULIS.

There are a few concepts in HULIS. Instead of beginning the house design with a room arrangement, HULIS begins by asking the consumer what kind of living he or she imagines, and then creates a design that can actualize the imagination. It is based on the notion that room arrangement constitutes the output of a lifestyle.

HULIS is made up of seven elements: (1) appearance, (2) structure, (3) entrance, (4) Japanese-style room, (5) Western-style room, (6) kitchen, (7) bathroom. The owner will begin with the appearance: "What kind of house do you want to live in?" If the image in mind is, "I want to live in a house with a gorgeous and tough appearance, and my budget is XX," then we input the Kansei words *gorgeous* and *tough*. HULIS has rules and a database for appearance, so it will generate the inference mechanism and display a graphic inferred from these Kansei words.

HULIS will then proceed to the entrance design: "How do you want the entrance to look?" If the answer is a *tough* and *massive* entrance, the HULIS computer will display an image of an entrance that looks tough. This continues through the last element, the bathroom. After the image for each home element is input using words, the overall house design is determined (Figure 5.1).

This Kansei engineering system (HULIS) will be a powerful weapon for an interior designer. Normally, a customer who wants to build a house has an image of what kind of house he wants to live in. HULIS is a system that can demonstrate the image realistically. By providing the HULIS result to the contractor, we are able to realize a result that satisfies all three parties—the user,

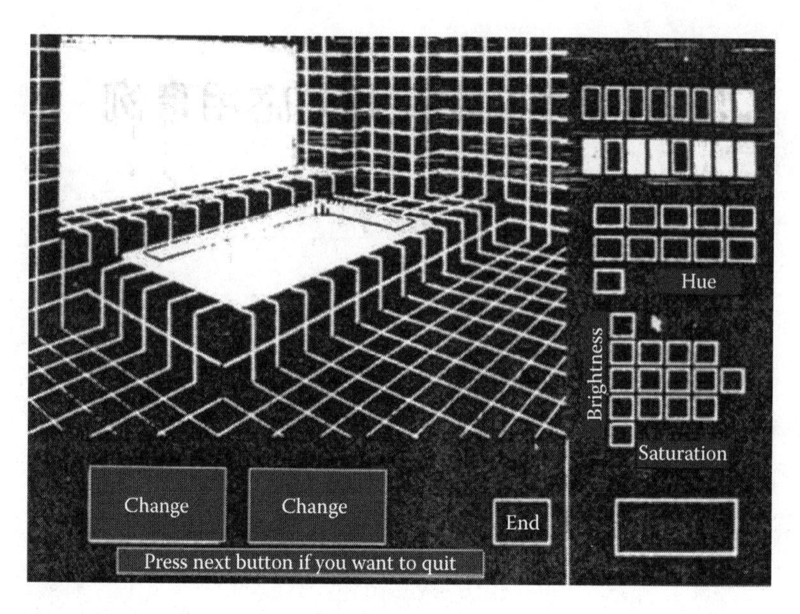

FIGURE 5.1
The output image (color) of bathroom from HULIS.

the architect, and the interior designer, who acts as the middle man. Actually, HULIS is a boon to the interior designer, who acts as an intermediary, taking care of the customer's image and influencing its realization (Figure 5.2).

If HULIS is developed to perform the following functions, it could become the second version, HULIS-II:

1. It will perform the inverted shadow calculation based on data of the land map brought by the customer and its surroundings, using software loaded into the computer. The land map is scanned, and the area will be automatically read. Next, based on the inverted shadow calculation rules (artificial intelligence), which are embedded into the computer, the system will calculate the effective three-dimensional space within which the house can be built.
2. Input the budget that the customer can allocate for the whole project. This will directly affect the quality of the building or available options.
3. The customer will be asked to talk about his lifestyle. The keywords are picked up and input into the system. For instance, "We are a family of four: my wife, our high school-age son, and our junior high school-age daughter, and me. My wife is also working in a company, and both of us are at the management level, so sometimes we come home a bit late. Saturday nights, we usually go out for dinner, but every one or two months we invite friends to come over for dinner and karaoke..."

FIGURE 5.2
Illustration of the use of HULIS.

From their living styles, annual income, and family structure, based on the embedded artificial intelligence rules, the system will infer how many rooms are needed for the children and how many rooms are needed overall, how the rooms should be arranged, how the entrance and guest room should be, how big the bathroom should be, and many other elements. The computer will propose a building structure such as the arrangement of rooms that matches the customer's lifestyle, and it will explain the reason as well. The customer will get a few proposals, and the proposals can be amended at the customer's request.

4. Back to HULIS, input images related to each room such as appearance, structure, entrance, Japanese-style room, Western-style room, kitchen, bathroom, and so on. The size and interior of each room will be inferred and determined using the Kansei engineering technique.
5. Consider the budget, perform the overall inference, display in color the architectural drawings (three views), room arrangement, and the Kansei of each room, together with the budget.
6. The cost calculation and detailed construction drawings will be printed, but of course these are not for the customer to see.

The customer will bring home the three-dimensional drawings generated in Step 5 and the family will discuss the contents. If everything is to the family's liking, the family will place an order.

By utilizing the inference using artificial intelligence, HULIS-II has bilateral merits allowing it to propose a design that can realize the lifestyle desired by the family, the family can have further discussions right away, and the family is able to participate in the design process.

Part of HULIS-II has been used by a construction firm, and it has been highly praised. We can get much more interesting results if we combine this Kansei engineering system with virtual reality. I will touch on this later.

5.2 FAIMS

FAIMS (fashion image system) is a computer system that uses Kansei engineering to decide the dress for female university students, as explained in Chapter 4. It is embedded with a database of all-season suits and one-piece outfits for female students. When the Kansei words describing the dress female students want is input, the clothes that suit their images will be displayed.

First, the computer will ask, "What image of dress do you want?" The user will input the words, like *cute dress* or *intellectual dress*. FAIMS will perform the inference by utilizing the Kansei words database and knowledge database, and then display the finalized dress design.

The computer will extract 50 types of dress design in one shot for the first word, but it will only show the highest-scoring option. If the user thinks the design does not fit her image, she can view other options, one after another—the second highest, the third, and so forth through the 50th. She can also change certain parts that do not suit her taste.

FAIMS can be expanded to the higher scale FAIMS-II. When we performed biometric measurement to 51 items of a dress using about 400 female university students, we found that all the measured values can be reproduced with three body measurements: height, chest, and iliac crest angle. By using these measured values, we can classify the female university students' body shapes into 10 types. In other words, we found that with three body measurements we can determine the body shape, and Kansei engineering can be applied to each shape as needed. We are able to create the image of *tall and cute* or *medium-built and cute* designs using Kansei engineering.

In FAIMS-II:

1. Height, chest, and iliac crest angle are measured. The computer will infer the person's body shape and turn it into an image.
2. The image of the desired dress is input using words. The computer will utilize the database and infer the dress that is logical to the Kansei word.

3. The inference result in Step 2 is fitted to the body shape inferred in Step 1.
4. The derived Kansei engineering graphic is displayed on the monitor and then verified as to whether it fits the customer's image. Partial alteration can be done on the computer.
5. If the customer is satisfied, FAIMS-II will output the pattern that matches the person's body shape and the desired image. The rest is just producing the actual dress.

FAIMS-II has been constructed only up to the step of identifying body shape and designing using the image. Integrating these two parts still remains an issue.

5.3 Entrance Door Kansei Engineering System

We have codeveloped the Kansei engineering system for entrance door design with Tateyama Aluminum Industry Co., Ltd. The objectives are to provide front doors that can satisfy customers who look for aluminum sash doors by asking them to express the Kansei that they have, and to build a support system that enables the R&D personnel to perform new product development appropriately and efficiently.

According to Kansei engineering procedures, the process started with extracting Kansei words related to the aluminum sash door. The marketing staff and designers cooperated in extracting the words. After extracting quite a large numbers of Kansei words, we held a discussion with the designers. This resulted in narrowing it down to 40 pairs of Kansei words.

The next step was collecting door samples, including the sash doors of other makers. We ensured that various designs were included, and we collected 82 slides of sash doors. We showed the slides to interior coordinators and the staff designers, a total of 77 people, and asked them to perform an evaluation using the aforementioned 40 pairs of Kansei words.

Among the identified basic shapes of aluminum sash door are those shown in Table 5.1. Additional considerations are the door's color and many other detailed characteristics. Table 5.2 shows some of them arranged into item/category related to 82 types of doors. The results of Kansei evaluation for each door were analyzed using quantification theory Type I, and then stored in a database for the Kansei engineering system.

The system structure for the entrance door Kansei engineering system (EDKES) is shown in Figure 5.3. Basically, it is almost the same as HULIS and FAIMS. The difference is the design database expresses the Kansei words database with HULIS, images database with FAIMS, and doors graphics

TABLE 5.1

Examples of Item/Category for Aluminum Sash Door

1. Frame type	(1) Single door (2) Double door (3) Double panel (4) Double leaf
2. Fanlight	(1) Door fanlight or Frame fanlight (2) No fanlight
3. Lattice	(1) Horizontal (2) Vertical (3) Crossed (4) No lattice
4. Door structure	(1) Top curve (2) Door curve (3) Top point (4) Flush (5) Others
5. Muntin	(1) No muntin (2) One piece (3) Two pieces (4) Three or more pieces
6. Glazing bar	(1) No glazing bar (2) One piece (3) Two pieces (4) Many pieces
7. Cross rail	(1) Have (2) Don't have
8. Double cross rail	(1) Have (2) Don't have
9. Clove	(1) Downward clove (2) Upward clove (3) Don't have

TABLE 5.2

Part of the Item/Category for Aluminum Sash of Entrance Door

Item	Category	Sample																								
		26	27	28	29	30	31	32	33	34	35	36	37	38	39	40	41	42	43	44	45	46	47	48	49	50
Frame Type	Single Door		●			●	●		●		●		●		●			●						●		●
	Double Door	●						●		●				●			●		●		●		●		●	
	Double Panel			●	●							●				●						●				
	Double Leaf																			●						
Fanlight	No Fanlight	●			●	●	●		●	●	●	●		●		●		●			●	●		●	●	
	Door Fanlight		●	●									●		●								●			●
	Frame Fanlight							●									●		●	●						
Door Color	White						●	●							●		●	●			●		●			
	Grey	●										●				●										
	Black									●				●								●				
	Pastel				●				●										●							
	Brownish		●	●		●					●		●							●				●	●	●
Lattice	No Lattice	●	●	●	●	●		●	●	●	●	●	●		●	●	●	●	●	●	●		●			●
	Horizontal																									
	Horiz. & Vert.													●								●			●	
	Crossed																									
	Others						●																	●		
Door Structure	Top Rail	●	●	●		●			●	●				●			●			●		●	●	●	●	●
	Top Curve							●			●		●		●											
	Door Curve																									
	Top Point						●																			
	F-oblong											●				●										
	F-others																									

with EDKES. When the Kansei word for the desired door is input to the EDKES that has been built up to process it, an image of the inferred door will be displayed on the computer.

EDKES can be set up and used in a marketing office, where a customer can use it to visualize the desired door. By putting together the assembly parts, we can set up the desired door that he wished for, giving the customer a sense

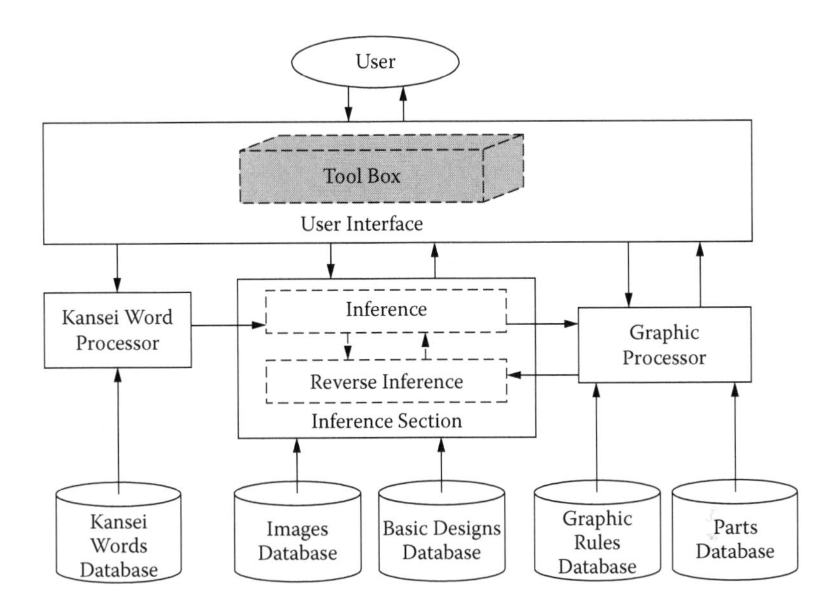

FIGURE 5.3
System structure for EKDES.

of satisfaction. Additionally, EDKES has been configured to automatically record the Kansei words chosen by the customers. This input will be helpful in enhancing the development process of entrance doors in the future.

Another interesting point about EDKES is that, by giving a specific Kansei word and the information on the doors' item/category, which is the attribute of the design, to a designer, a totally new product can be produced. In one example, the attributes (design elements) of an entrance door for the Kansei *simple* door were as follows:

1. Door structure—Vertically long flush
2. Frame type—Single door
3. Door color—Gray
4. Muntin—Two pieces

The designer used this information and a short while later produced a drawing. After experimenting with many other Kansei words, we got many new door designs. A subsequent interview with the designer revealed that if they get this kind of information, they don't have to take the trouble to think about so many things. While visualizing the Kansei word's image, the door design attributes pop up in a designer's head with almost no effort. The designers proved that such an approach is a very effective development technique.

5.4 Kansei Engineering System for Car Interior

Many types of software have been developed in Kansei engineering. Another kind of extraordinary application is in passenger car design, for which we have developed a few Kansei engineering systems that are being utilized. KEES-D (Kansei engineering expert system-D), which I am going to describe here, is a design support system tailored for car interior design.

I have mentioned that Kansei engineering is a technology that translates Kansei words like *elegant, beautiful, luxurious,* and so on to product design, and I have explained the application in detail. What I am going to explain now refers to car interiors, but it is more specific to *spacious feeling, relaxed feeling, tightness feeling, narrowness feeling,* and the like—the kind of comfort related to the size of the interior. In this sense, it is a little bit different in terms of characteristics compared with the Kansei that I have mentioned previously.

There is a constraint to the size of a car. A design can be praised as good if it can psychologically give the sense of *spacious* to the driver and other passengers even with a size constraint. For cars that provide satisfaction through speed, like a sports car, a good design is one that can make the traits of a sports car be felt more strongly by creating a sense of narrowness rather than spacious. However, in order to materialize the Kansei such as *spacious feeling* into the design, we need to know the interior factors that contribute to the spacious feeling. This can be discovered using Kansei engineering. Furthermore, by utilizing the result, it is also possible to build a computer system that can realize or diagnose the spacious feeling. KEES-D is a Kansei engineering system that was researched and developed for these two objectives.

We started with the initial objective, which was to analyze the Kansei of a car interior. The research began with the extraction of Kansei words. A few hundred words were collected, and with the help of designers as car professionals, 100 adjectives that are logical to use to describe a car interior Kansei were selected and given their respective antonyms so that they could be used in the form of the SD scale. We prepared a 5-level rating scale (Figure 5.4).

Next, since the target was small-sized passenger cars, we prepared 20 small-sized cars with 1000–1500 cc engines, and had them evaluated using the aforementioned 100 Kansei words on the SD scale. The evaluators were individuals involved in car manufacturing, with a total of 41 persons of both genders. This is how we got the Kansei data for the small-sized car interior.

On the other hand, regarding the car design, we promoted the design elements identification process—what has been referred to as the item/category classification process. The car designers must put a lot of effort into this process. Therefore, we briefed them in advance about Kansei viewpoints and how to determine the item/category. This is the most important point

FIGURE 5.4
Car Kansei engineering.

in Kansei engineering. If there is a mistake in controlling how people classify the thing they see, we won't get a good result during the statistical calculation. In this research, we obtained 224 types of item/category for the car interior design.

The computer architecture of KEES-D was built as in Figure 5.5. This Kansei engineering system itself has software that analyzes quantification theory Type I and a checker that performs a dependency check before the analysis. These points are fundamentally different from the existing Kansei engineering systems. The dependency check means all the items are checked to see if they can be handled independently or not (i.e., if overlapping occurs at some points between items, it is called dependency), and if statistical analysis is possible or not.

First, we applied the quantification theory Type I to the design evaluation result using the 100 Kansei words to grasp the relativity between the image and item/category. By checking the data, we confirmed that *spacious feeling*, *relaxed feeling*, *tightness feeling*, and *narrowness feeling* relate greatly to the interior dimension. Hence, we extracted only the data related to these four Kansei phrases and the interior dimension data, and then constructed a database system as in Figure 5.5.

A designer who wants to examine the *spacious feeling* of a small-sized passenger car using only the interior dimensions will sit down in front of a computer with KEES-D installed and run the KEES-D. Initially, the KEES-D will show a display and ask the designer to input the dimensions. In this example,

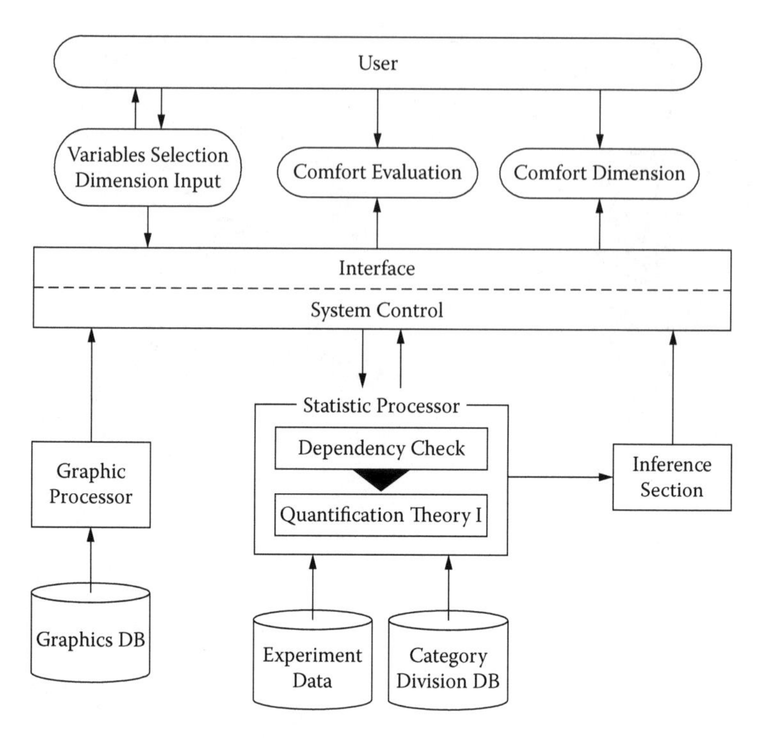

FIGURE 5.5
Schematic diagram of KEES-D.

they are XL 38 (the distance from the driver's eye to the bottom end of the left pillar) and W 20 (the horizontal width from the eye to the car's center line), and the predetermined dimensions are input. Next, a few diagrams will be displayed one by one, and dimensions are input for each diagram.

When all the inputs are completed, the KEES-D statistic processor, as shown in Figure 5.5, will execute. First, it will perform the dependency check, then recall the experiment data and perform quantification theory Type I (Figure 5.6). After the specified process is performed, based on rules defined in the inference section, the spacious feeling will be evaluated using the input dimensions.

In this case, the score for *spacious feeling* is 65 points, while the score for *tightness feeling* is 63 points. This tells us that the interior dimensions give a certain degree of feeling for both spaciousness and tightness. The KEES-D also has a description feature that explains how it gets the score.

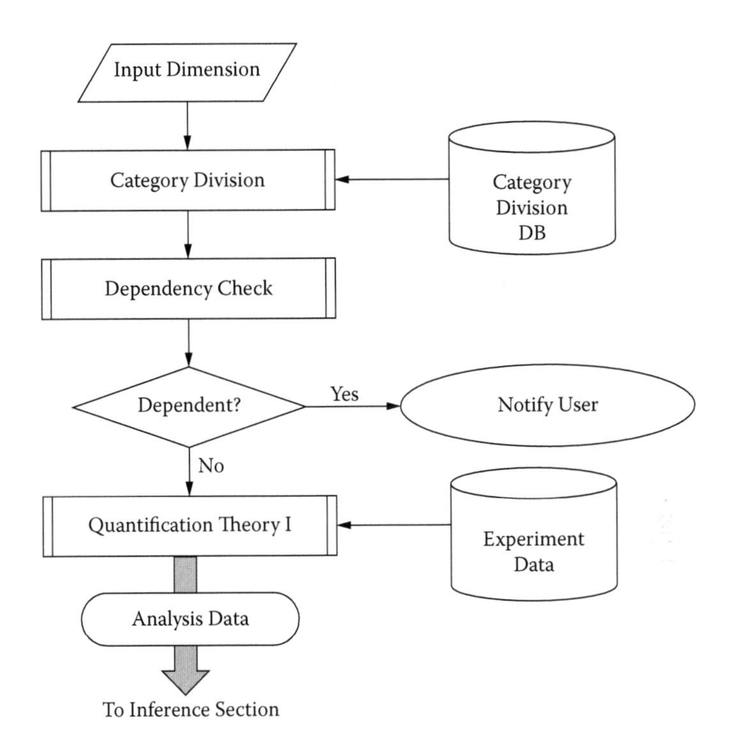

FIGURE 5.6
Flowchart of the statistical processor.

6

Hybrid Kansei Engineering

6.1 Forward Kansei Engineering and Backward Kansei Engineering

I have explained that Kansei engineering is a technology that translates the Kansei, or images, that consumers have in their minds regarding product designs. Kansei engineering using this approach is called forward Kansei engineering. There are two methods in hybrid Kansei engineering, forward and backward.

6.1.1 Forward Kansei Engineering

6.1.1.1 Consumer Product Selection

Forward Kansei engineering is a method whereby consumers select the product that matches their Kansei, as indicated by the Kansei that the consumers have in their minds. Then, a computer is used to display the corresponding design. FAIMS and HULIS fall under this category.

In the case of forward Kansei engineering, the targeted products are evaluated from the consumers' viewpoint using Kansei words. By utilizing the database, the relationships between the Kansei words and the item/category are decided. However, consumers obviously are not experts in product development. Thus, their senses related to Kansei are not precise. Therefore, for complex products or those that require specific higher-degree Kansei, evaluation data by the designers are sometimes needed in the databases.

6.1.1.2 Designer Product Development

Another way of using forward Kansei engineering is for designers or R&D personnel to be clear on the product design process. In such cases, the evaluation data by designers is used to build up the database. However, there are cases where the evaluation data from consumers are used as references.

FIGURE 6.1
Hybrid Kansei engineering system.

When a designer decides that he wants to design a product with "XX Kansei," he will input "XX Kansei" into the forward Kansei engineering system. The knowledge base will work to effectively support the designer's product development (Figure 6.1).

6.1.2 Backward Kansei Engineering

Actually, there is a method to support designers in product development more effectively. It is called backward Kansei engineering.

If we look at Figure 6.2, there are arrows in two directions; one to the right and the other to the left. The upper arrows show the forward Kansei engineering system explained earlier. It is the direction in which Kansei information is input from the left side, translated to the design elements, and then to the actual designing of a new product.

However, surely the R&D personnel and designers want to know whether the product that they picture in their minds matches their image of Kansei. If possible, they probably want to know what kind of Kansei eventually is implied by the product—whether it is close to or very different from their image. This is made possible with the backward Kansei engineering system, shown by the lower arrows in Figure 6.2.

In the backward Kansei engineering system, the drawings (sketches) are loaded into the computer. From the loaded image, the computer will recognize the shapes and colors and verify them. Then, the knowledge base of the Kansei engineering system will capture the image recognition result

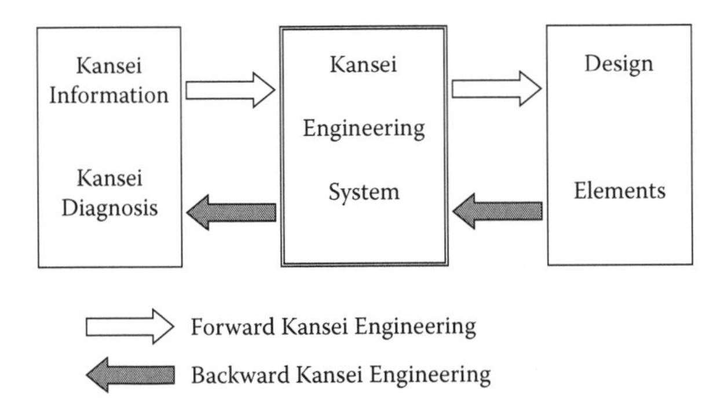

FIGURE 6.2
Hybrid Kansei engineering.

in the reverse direction. The system will then perform the Kansei diagnosis by inferring the most logical Kansei for the drawing pattern.

6.2 Hybrid Kansei Engineering

The system that has both the forward Kansei engineering system and the backward Kansei engineering system embedded into a computer is called the hybrid Kansei engineering system.

The advantages of the hybrid Kansei engineering system are as follows:

1. Designers can decide the product concept of the new product, input it into the forward Kansei engineering system, and get a rough idea for the design.
2. Designers can add their own creativity in addition to the guide, sketch the product based on the idea, and then use the backward Kansei engineering system to know the kind of Kansei and the specifications for the drawings.
3. By using the forward and backward systems consecutively, designers are able to design a product that perfectly matches the Kansei along with their added creativity.

We have accumulated considerable experience and built various databases as well as knowledge bases for the forward Kansei engineering system of the hybrid Kansei engineering system. Thus, we have developed know-how techniques. The difficult part is the backward Kansei engineering system.

Ideally, in the backward Kansei engineering system, a designer can just input a pencil-drawn sketch, and the system will read and diagnose the corresponding Kansei and its specifications. However, in order to achieve this, the backward Kansei engineering system must have a common sense capability so that it is able to recognize images at almost the same level as humans can. Teaching common sense to a computer cannot be accomplished without great effort.

One possible way to realize this idea is by specifying a product field, that is, a product domain. By doing so, shape, color, function, and so forth are specifically determined, so there is no need for the knowledge to be as high a degree as common sense. With this, we can actually build the capability that was thought to be almost impossible into a simple form. In other words, by embedding a simple intellectual technique into a computer, it is possible to have a recognition mechanism.

Here, I would like to introduce techniques that make shape recognition easy in the backward Kansei engineering system. One such method is a menu-driven technique. In the forward Kansei engineering system, the item/category, that is, the product's physical characteristics that match the input Kansei, are determined through inference, integrated, and displayed on the monitor in the form of an image. The backward technique reverses this process. In other words, the item and its detailed category are extracted from the component database, which the designers already have, and they will be asked to choose the product in mind from a menu. The structure of the system (menu-backward Kansei engineering) is shown in Figure 6.3.

When a designer selects one option from the menu, the computer will configure the whole image from the component and display a drawing. When the designer approves it, the inference mechanism is activated to diagnose the Kansei of the drawing by utilizing the knowledge database and Kansei database. However, we still have an issue here: The inference from Kansei to the component and vice versa does not result in a one-to-one pair. Forward translation and backward translation do not always tally (there can be a mismatch of mapping and reverse mapping). However, we thought of a few information processing formulas to improve the chances for success.

6.3 Backward Kansei Engineering with Template

An advanced method in the recognition of the input image is the technique where the computer has templates as samples and performs shape recognition by matching the input information to those templates.

The conditions in performing this technique are as follows:

1. The computer must have templates for all the components that could be enclosed in the anticipated input images.
2. The computer would not recognize if a designer carelessly gave image information that is not relevant to the template.

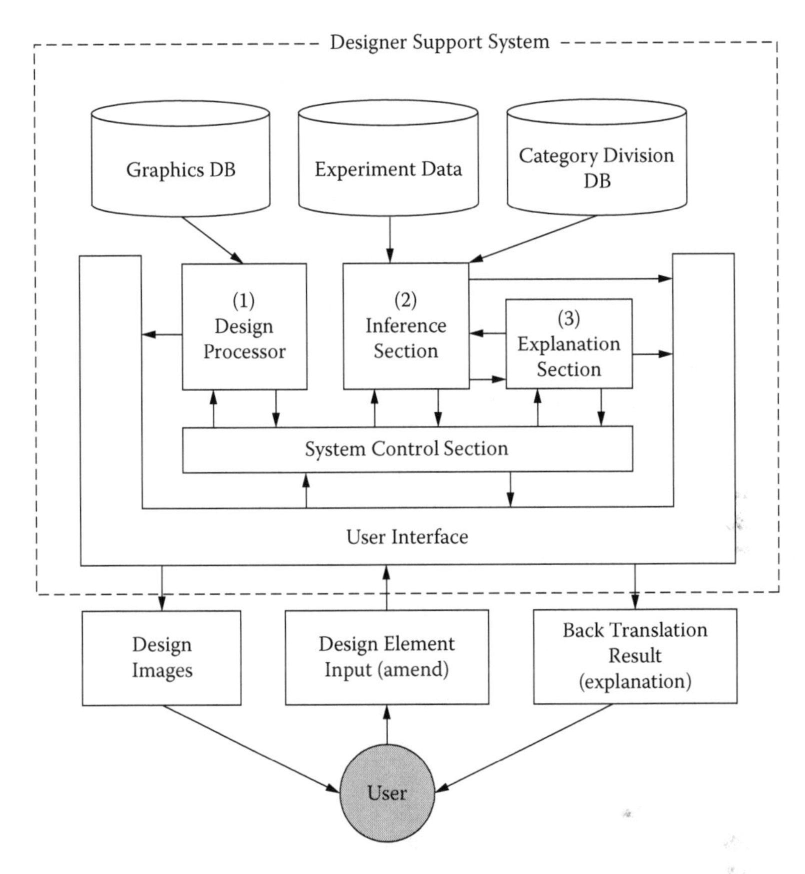

FIGURE 6.3
Menu-backward Kansei engineering system (MBKES).

When the anticipated input components are grasped in advance, this template technique can be used to perform recognition very efficiently. However, this system is not suitable for complicated three-dimensional images or uncontrolled images where templates cannot be prepared.

I will introduce the actual application case of the template recognition system. Previously, I introduced an actual success story in applying Kansei engineering to an aluminum sash entrance door. We have actually developed a template-method backward Kansei engineering system related to this.

6.3.1 Image Recognition Procedure

The backward translation system will perform image recognition for the input image information through the following steps:

1. Image conversion—Based on the shading of the input entrance door image, the system will perform the gray value conversion and

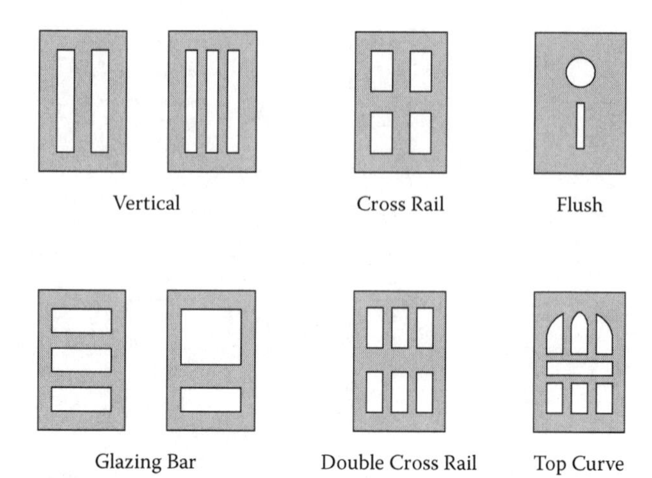

FIGURE 6.4
Recognition of each door.

recognize the outline (outer frame) of the entrance door. Once the outer frame has been identified, it will then proceed to recognizing the presence of a fanlight and the pattern of the entrance door.

2. Characteristics extraction—Next, it will go into recognition of contents of the entrance door by analyzing lines, areas, and angles. Using the recognition of angles and lines, as well as the judgment of areas based on the angles and lines, the system will recognize the presence of a fanlight, the pattern of the entrance door, and so on.
3. Pattern recognition—The elements that show most of the entrance door characteristics are the vertical and horizontal lines, as well as the presence of lattices and flushes. The computer has the standard patterns, and by overlapping (matching) them with the input image, it will perform pattern recognition, such as recognition and counting of the vertical and horizontal lines (Figure 6.4).
4. Color recognition—The colors of the input image are recognized using R×G×B calculation, and they are translated into the Kansei of colors.

6.3.2 Backward Kansei Engineering System

The structure of the backward Kansei engineering system for the entrance door is shown in Figure 6.5.

The input image will go through the image recognition process as above before performing the outer frame recognition and the door pattern recognition such as *single door* and so on. Then, the computer will ask the user to confirm the design. Here, the result of the image recognition is displayed on

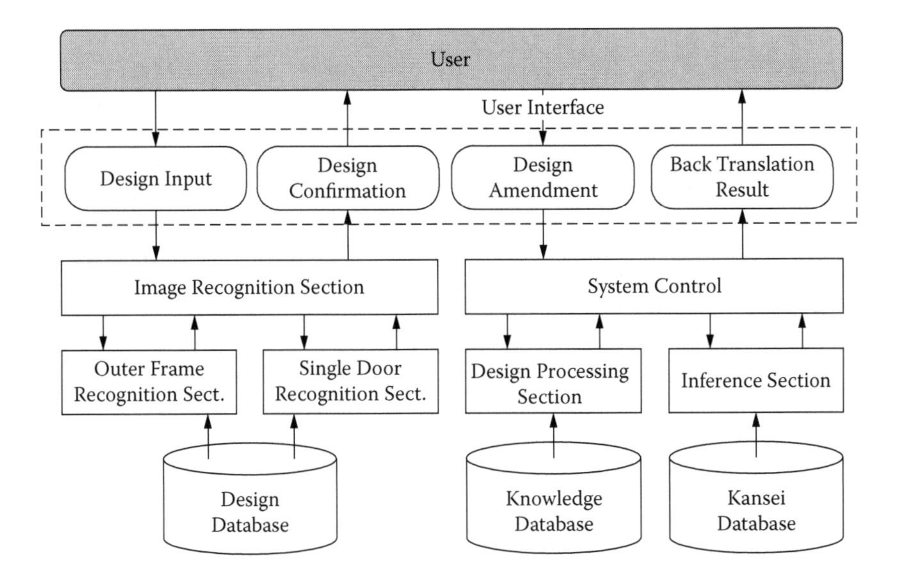

FIGURE 6.5
The system structure.

the screen. At this stage, the system will utilize the knowledge database and Kansei database to complete the Kansei recognition.

The designer may wish to amend the displayed recognition image, so recognition of the amended image begins, and the system performs the Kansei diagnosis by executing the back translation using inference.

6.3.3 Actual Example of the Recognition Result

An example of the back translation result using the template method is shown in Figure 6.6. On the upper right is the input drawing of a double-opening door. The drawing read by the computer's image recognition system is the double-opening door shown in the middle. We can see that all the characteristics of the image have been recognized. The list on the left side is the contents recognized by the image recognition system shown in the format of item/category. The Kansei diagnosis result for the case is displayed using Inference Technique IV. At present, it is not clear which inference technique is the most appropriate for the following three points:

1. Openness
2. Unconstrained
3. Cleanliness

All three have been inferred by several techniques, and they match the predicted content using the forward Kansei engineering system. Therefore, we

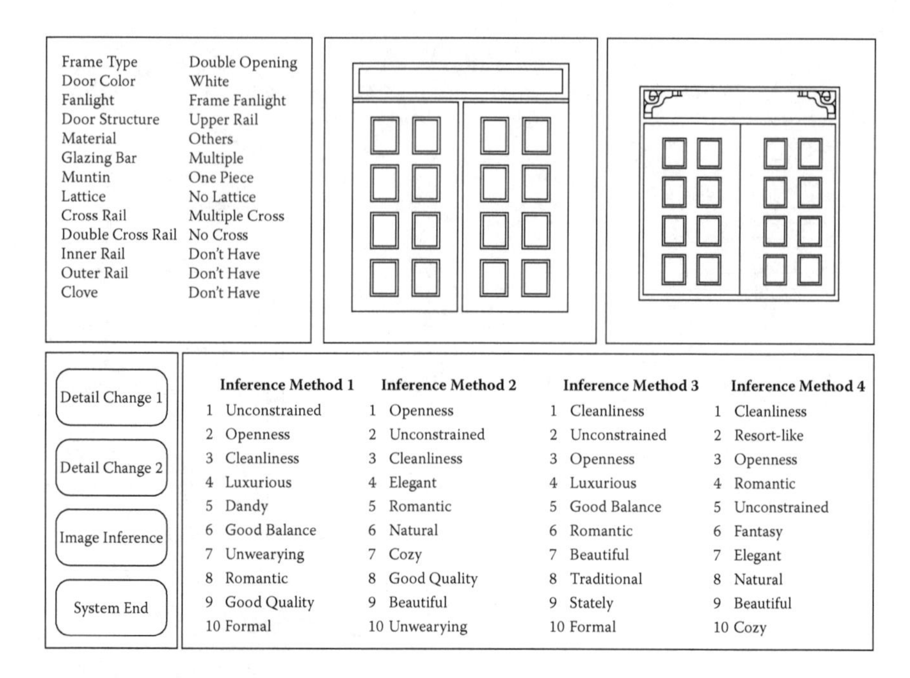

FIGURE 6.6
Output example (recognition result).

can conclude that hybrid Kansei engineering is possible for images that do not have very complicated structures.

6.3.4 The Future of Hybrid Kansei Engineering

Hybrid Kansei engineering refers to a system where the mapping from Kansei to product design elements is possible, and at the same time, the reverse-mapping from the designed product to Kansei is also possible. It is a very difficult task to match the mapping and reverse-mapping results. Even though it is desirable to have a perfect match between both items, it is not problematic if they do not match perfectly. From its origin, Kansei has had the characteristic of *ambiguity*. If the sides roughly match and they are reasonable, it is acceptable.

In fact, Kansei engineering is not a technology created to replace designers. It has the role and function of assisting in roughly narrowing down the task, so that the R&D people and designers don't have to worry about producing an ineffective design. By starting from a conclusion provided by the Kansei engineering system and combining it with designers' creativity, a product that is unique to the person can be developed. The system that supports the starting point of this is the Kansei engineering system. It is not a system that produces even a rough conclusion. The same goes for the backward Kansei engineering system of hybrid Kansei engineering, where it needs a technological depth.

At this moment, research has progressed to a point where the backward Kansei engineering system is possible by limiting the product domain. When a designer inputs the shape drawing or captures the picture of a handmade product with a scanner, the computer will extract the outline based on shades and other information, and then perform shape recognition. Currently, recognition is performed using the characteristic extraction of a clear shape obtained through shade processing using a wavelet function. At the same time, as for the color, RGB information processing is possible, and we can extract color information. From these two, we have made image recognition possible. Since most of the products taken as subjects in Kansei engineering are three-dimensional, if we perform multilevel information processing between the three-dimensional processing, partially or as a whole, quite a detailed recognition will be possible.

When such a system is completed, the work of designers involved in the product development will become more efficient, with most portions of the work becoming computer aided. Thus, designers will be able to practice a more effective role by focusing on their creative work. Additionally, if this kind of hybrid Kansei engineering is used in education, it will be possible to train good designers.

7

Virtual Kansei Engineering: Kansei Engineering Type IV

7.1 What Is Virtual Reality?

We often see the term *virtual reality* in newspapers and magazines. Virtual reality refers to a computer-generated virtual world that is very close to reality, and simulated experiences that closely resemble the real world.

Computer hardware has evolved tremendously and become a lot less expensive. This has made possible the creation of the three-dimensional virtual world, and the virtual, simulated experience in computers. In virtual reality, we are able to act freely of our own will, and this is basically different from dreams that are beyond our control.

The easiest way to demonstrate an example of virtual reality is with the flight simulator for pilot training. The outside world visible to the pilot is the approximate image of a runway and other items that are created using computer graphics, and the image will respond to the operational changes in the cockpit. With the complex operation of hydraulic devices, gravity changes that are similar to changes in reality are applied to the cockpit to give the feeling of takeoff and landing. All these are being controlled from a computer. With the flight simulator, which has been developed using state-of-the-art technologies, there is an advantage of being able to greatly reduce the number of field trainings. In fact, quite a big portion of pilot training is now done using the simulator.

The main reason simulators are used for pilot trainings is that aircraft are an expensive commodity, and accidents caused by inexperienced pilots might result in heavy losses, including human lives. Another reason is that besides being able to simulate the rule-based flight using computers, training for emergency situations is also made possible by simulating emergency circumstances.

Why is virtual reality good? Compared with actually creating a real object or situation, creating it as a computer image is cheaper. On top of that, since these are computer images, there are no repair and maintenance costs. Also,

FIGURE 7.1
Virtual reality and Kansei engineering.

by utilizing the simulated experience, the human experience of a situation using computer technology is like a walk-through of the actual situation.

In the virtual reality field, many companies are involved in developing hardware and software, which include a computer to handle images and software, an HMD (head-mounted display) to show three-dimensional images in front of the user's eyes, data gloves to hold or to operate the items in the images, and a tool for image creation (Figure 7.1).

Basically, the graphics created using a tool for computer graphics are alternately projected to the left and right displays of the HMD in a slightly shifted form. Users can blend into the image, which is in a three-dimensional space, and with hand movements over the data glove, simulated experiences such as catching something or turning on a faucet in the virtual world are possible.

7.2 Virtual Kansei Engineering

Virtual Kansei engineering is a technology that integrates virtual reality and Kansei engineering. Hiroshima University and Matsushita Electric Works Ltd. (MEW) did joint research to develop virtual Kansei engineering as a new technology—the first in the world to combine virtual reality and Kansei engineering.

The virtual Kansei engineering that we developed is a system that provides products that can satisfy customers' needs in purchasing a MEW custom kitchen. First, the customers will describe the lifestyle they dream of. Keywords from the description will be input into the virtual Kansei engineering system. Next, they will be asked to describe the image of the kitchen they want in the form of phrases like *elegant, a bit luxurious,* and *a convenient kitchen.* Kansei words from this phrase will be input into the system. Automatically, the picture of a custom kitchen close to the image described will be displayed on the computer. If customers are roughly satisfied with that, they will step into the virtual reality, put on the HMD and data glove, and enter the custom kitchen that the computer has selected based on the Kansei. Customers can then examine the kitchen by trying to cook, touching appliances with their hands, and turning on the faucet using the data glove. Customers also check the locations of the cupboards, dishwasher, and so on to their satisfaction. Anything unsatisfactory can be amended in the computer.

When a customer is fully satisfied, the design drawing of the custom kitchen will be output and transmitted to the MEW factory. The factory already has its manufacturing system equipped with computer-integrated manufacturing, so the drawing will be incorporated into the production plan, and within one to two weeks' time, the product will be manufactured and delivered to the customer. In this manner, the custom kitchen, which is based on the customer's Kansei and has been virtually experienced by the customer to his or her satisfaction, is manufactured in a short time and will give full customer satisfaction. The particular objective of virtual Kansei engineering to develop products aimed at customer satisfaction through customer participation. In MEW, this is called the *ViVA system.*

7.3 Custom Kitchen Kansei Engineering

The outline of custom kitchen Kansei engineering is as follows.

7.3.1 Computer Memory Content

The first thing that the computer system must know is the customer's kitchen area. The kitchen layout is input in the computer: length and width; position of pillars, doors, and windows; the style of kitchen space; and so on. The computer will separately store in memory the sink, dishwasher, cabinets, and so forth as components.

7.3.2 Selecting Kitchen Style

The customer describes his desired lifestyle in terms of the kitchen. The keywords identified from the description are arranged and inferred based on the rules acquired from experienced kitchen designers, and the style of the custom kitchen is selected. In the rules that the designers have, the kitchen layout is almost completely determined by the income, age, cultural characteristics, and other characteristics of the customer, such as lifestyle. The person's lifestyle especially has a high correlation to the kitchen layout. These kinds of rules are embedded into the knowledge base.

In the next step, the customer describes the image (Kansei) that he has regarding the kitchen. This image will determine the design style of the kitchen. Additionally, the cupboard design, cabinets, wallpaper, and color are also determined.

7.3.3 Virtual Reality

When the computer has determined the design of the custom kitchen, the images are sent to the virtual reality system. By putting on the HMD and the data glove, customers can blend into the computer image. They can move around freely in the system, touch the desired areas, turn on the faucet located at the sink to run water, and check the height of the countertop beside the sink by doing some cutting. They can even open the upper cabinet doors to confirm the cabinet height.

With virtual Kansei engineering, by virtually experiencing everything inside the simulation of our own image, we can confirm the compatibility of our own Kansei product and its usability by direct experience. It is a technology in which the customers participate in the design process to directly achieve their own satisfaction. Such a system that realizes customer satisfaction in combination with virtual reality will probably become standard in future product development.

MEW has recently developed a system where the whole family can share and enjoy the virtual reality. In conventional virtual reality, the simulated experience was not possible without putting on the HMD and data glove. Therefore, only one person at a time could experience it. However, for cases that involve the whole family, like custom kitchen and house design, the design satisfaction through the participation of all members is essential. Based on this point of view, MEW has developed a system that enables many people to participate in one virtual reality. This technology can genuinely be called a participatory design system.

7.4 The Evolution of Virtual Kansei Engineering

Virtual Kansei engineering is being positioned as Kansei engineering Type IV. I can imagine how this new technology can be applied and how it will evolve in various industries. The following applications and evolutions are considerations.

7.4.1 Fields Where Trial Products Are Either Expensive or Require a Long Design Time

The original intention of using virtual reality was to try an actual simulation on a computer to study its competency. It is about building a virtual world such as outer space, an ocean bed, a body's interior, or a town that is going to be designed and built, and examining various issues inside those virtual worlds.

In the field of Kansei engineering, it would be very expensive to actually build a house and conduct testing and studies. Additionally, it is not easy to modify something on an actual building. However, building images in a computer is possible if we are willing to invest some time. As described earlier, since we can execute our own actions, such as opening and closing a door in the virtual world, it is worthwhile to consider virtual reality technology.

House design, custom kitchens, and others mentioned earlier are fields that could utilize this technology in the future. From our viewpoint, since we have already completed the HULIS for housing, it won't be that difficult. Additionally, this technology can also be utilized for passenger car exterior and interior design.

The R&D personnel could go into the virtual world, open a car door, settle into the seat, and check the interior design by touching the steering wheel, audio equipment, and so on, or even check the speed and maneuverability by actually turning on the engine. All these capabilities could be realized in the near future.

Virtual Kansei engineering is also expected to be used in town planning or landscape design, as well as in examining the exterior and interior design of buildings.

7.4.2 Customer Decision-Making

As something that provides a realistic sensory experience in addition to the existing Kansei engineering, virtual Kansei engineering can be utilized as a decision support system to assist customers in product selection.

For example, we could upgrade the FAIMS and develop a Kansei engineering system for a three-dimensional view of fashion. Then, we could input the customer's Kansei to enable the system to display the corresponding image. Another camera could be used to recognize the characteristics of a customer's face. The system could perform biometric calculations, and the customer could put on a dress newly designed according to her image. She can see herself in a three-dimensional space. Any actions she takes would appear in the virtual world. An appearance check can be performed by using this technique as well.

The system that will be described here is not an authentic virtual reality, but we can still use it. For a passenger car, the exterior and interior images could be loaded into a laptop computer, and the salesperson could bring the computer to the customer's home and show the images to all the family members. Using the walk-through technique, they can view the appearance and interior of the cars. This is a tool that can be used in decision-making.

From this point of view, it is possible to build a system to display virtual reality in a portable computer by using Kansei engineering. In the near future, perhaps it will be a trend for salespeople to carry laptop computers loaded with databases. At each customer's home, the salesman would identify the customer's Kansei and input them into the computer. Images would then be displayed, and the customer would be asked to walk through the images and confirm the product. I can imagine such a scenario, and I believe that this one, too, will be realized in the near future.

7.4.3 Education and Training

In the operation of large-scale systems like chemical plants and power plants, a network of advanced systems such as artificial intelligence is set up to help control the system in serving its purpose. However, the operations are being performed by humans, who are sometimes unable to keep pace with technology. Thus, there is a risk of human error. The effects of such error could be massive due to the size of the plant.

When we studied the causes of human error, we found that there are two big factors involved in humans: One is the human's cognition problem, and the other is the emotion problem. As for the former, the larger the scale of the system, the more knowledge and skill required for operation. This cannot be fully covered with specialized training. The possibility of encountering unusual situations during training is minimal. The hidden portion is huge, and this leads to human error due to lack of cognition.

For the latter, emotion, the system is even strengthened intellectually by including artificial intelligence, as the operators are still human. The sudden change of situation would affect the emotion of the operator, causing an emotionally stressed operator to perform incorrect operations, resulting in a human error.

Either way, cognition and emotion are the two primary sources of human error, especially in a large-scale system. In considering these two points in Kansei engineering, and by combining them with virtual reality technology, it is possible to utilize virtual Kansei engineering in developing human error prevention training. This Kansei engineering technique can also be utilized as a diagnosis system for human error occurrence by incorporating it into the large-scale system.

We conducted joint research with a power company for utilizing virtual Kansei engineering for nuclear power plant operating system training. We built an expert system on the emergency procedure guides, and in that system we included a human model that can differentiate and recognize human error occurrences as either cognition or emotion. On the other hand, for virtual reality, in order to enable the power plant operation, we built a three-dimensional panel using images, and the operators can virtually touch the panel.

Then, we simulate an abnormal situation, and the operator will perform the appropriate operation. The human model based on Kansei engineering will analyze the movement and guide the operator to take the appropriate action. Whenever necessary, the reactor condition and auxiliary equipment condition are shown as images in a window to enhance the operator's understanding. In this manner, it is possible to give the operators training that combines the operation and the simulation of an actual situation. Kansei engineering is one kind of human model, and it is also possible to use it as an education and training model.

8

Kansei Quality Management

8.1 What Is Kansei Quality Management?

Kansei quality management is a relatively unfamiliar term that was first proposed in the Union of Japanese Scientists and Engineers' meeting a few years ago. Kansei engineering Type V actually refers to Kansei quality management. Kansei quality management is described as the application of quality management that begins with the customer's Kansei, with the aim of maximizing customer satisfaction.

Japan's economic level has reached the world's top echelon. Many factors have contributed to that, but the biggest is probably the cultivation of the capability to produce inexpensive yet good-quality products. In fact, the idea of quality control (QC) was developed in Japan, as was total quality control (TQC), which was developed from QC.

QC refers to a control technique to maintain product quality at a constant level. It is well known that Japanese companies learned the philosophy and method of statistical quality control (SQC) from Professor William Edwards Deming, the post–World War II American statistician, and applied it in the production plant. Statistical quality control is a technique for determining the factors necessary for keeping the level of quality within a constant range by analyzing the data related to quality variances using statistics. With this technique as a foundation, the movement to ensure the quality of a product, beginning from the shop floor, has been performed in the form of small group activities. This has had a big impact on the quality improvement of products in the market.

The QC movement activity contributed to quality improvement in many companies. This improvement activity eventually became popular and was introduced all over the world. It later began to be called *kaizen*.

On the other hand, there was the suggestion that in order to ensure product quality, a company-wide movement needed to be instituted. This led to quality assurance to customers, and the movement to establish a company-wide quality assurance system began. This is known as TQC, which adopts an approach in which the top management target is broken down until it reaches the lower levels of the organization, and based on that, quality improvement is performed.

In Japan, the Deming Prize, named in honor of Professor Deming, has been established to reward companies that have introduced TQC, made efforts to improve, and achieved excellent results. The prize is awarded annually.

QC, which was originally learned from America, and TQC, which was created and enhanced in Japan with unique characteristics, are now being adopted by America and Europe and have been introduced in many companies in many countries. Both have produced good-quality products throughout the world.

The University of Michigan has a program called the Japan Technology Management Program (JTMP). The JTMP assists American industry in learning from Japanese approaches to the management of technology.

I have cooperated in the project on a Japanese committee. If we compare major American companies and major Japanese companies, there is almost no difference in terms of quality management, production system, and even new product development technology.

Regarding TQC, America has a broader viewpoint of management methodology called TQM (total quality management). Japan's Deming Prize puts emphasis on the policy and its expansion, as well as its standardization. On the other hand, America's Malcolm Baldrige Award evaluates on a broader basis: the top management's leadership, quality, and process management, as well as customer focus and social contribution.

8.2 What Are the Aims of QC and TQC?

If QC and TQC flavored using Japan's unique recipe are totally adopted and introduced in America and Europe, in the near future, Japan's companies will lose their competitiveness. For example, Ford's former CEO Donald Petersen wrote in his book, *Teamwork*, "We learnt QC from Mazda, and we learnt Toyota Production System from Mazda and Toyota. We used QC movement in conducting the Board of Directors meetings, and the decision-making became impressively fast. Next, we introduced QC movement to the managers meetings. We also introduced Toyota Production System and QC methodology to shop floors, and we were able to make cars very efficiently and at low cost. On top of that, we learnt Kansei Engineering from Mazda, and we were able to develop the well accepted car, Taurus. When we come to this level, Japan is no longer our rival." Therefore, Japan must take immediate action to reinvent its original management and technologies.

So what is the aim of QC? It is to standardize the control management in order to maintain a high level of quality. It is to establish a system where products of the same quality can be produced regardless of who makes them. As a result, customers who buy the products will be happy. In this sense, the aim of QC is to achieve customer satisfaction. The customer will be happy because the quality is improved and the product lasts a long time. This will result in the improvement of the company's profit in the long run.

However, if the standardization of quality is implemented without any continuous change, customer satisfaction will decline. Demand for improvement on the quality level will emerge naturally. If we forever maintain the same quality level, it will someday become obsolete, and the customer will get bored with it and come to dislike it. If quality control is one of the factors in achieving customer satisfaction, then quality improvement must be performed continuously. When quality improvement reaches its peak, we could change its vector. In certain cases, we may add new functions or embark on a product evolution. Here, another new product development surfaces, and it will follow the established circle of quality improvement. In other words, breakthrough is always required in QC.

Even for TQC, quality improvement is performed based on the top management's policy. After the standardization of various control items is set, if it is followed forever, the company will become stagnant and stop advancing. All staff of the organization become stereotyped and will do only what has been standardized, rather than reacting to the changes of the times, and the company becomes excluded from the vector of customer satisfaction. Additionally, the stronger the promotion of TQC, the tougher the organizational control will be, and the unified force will act strongly, disabling new ideas within the organization.

The management technique required in the latter half of the 20th century was actually based on the philosophy of an agile organization. It meant becoming a corporate organization that has predictive and agile sensitivity toward the changes of the times and is able to respond to those changes promptly. For example, during Japan's bubble economy, what was the management technique and practice of Japan's companies? And when the bubble economy collapsed, what kind of agile response could be made?

TQC activity has made a big contribution to Japan's economy. It developed the capability that enables high quality and low cost management. However, at the same time, conventional TQC has made companies lose their thinking power, inculcated philosophies that vanished with the bubble, and caused companies to lose their ability to act as agile organizations. In other words, it has caused companies to neglect the changes in their surroundings and the realization of customer satisfaction. The TQC that was supposed to ensure customer satisfaction, but ended up with less committed penetration, turned them into organizations that neglected their customers.

It is time to think seriously about what customer satisfaction is and what we should do about it. The quality management technique that focuses on customer satisfaction is called Kansei quality management.

8.3 Kansei Quality Management Thinking

Kansei quality management is customer-focused quality management. It is management that grasps the customer's Kansei and improves the quality level to always satisfy the customer's Kansei (Figure 8.1).

Even if ambiguous, a customer surely has certain Kansei when approaching a company and making a request of the company in the form of an expectation based on Kansei. For example, a customer who wants to purchase a car will visit a car dealer with a general image in mind. There will be many Kansei requirements in terms of car design, functions, price, and so forth contained in the customer's Kansei. Of course, the dealer's attitude and how he treats the customer are also influencing factors. Even if there was a car that fit the customer's Kansei, if the customer service does not meet Kansei requirements, the customer might go to other dealers. A customer might even switch to another automobile maker. This theory is

FIGURE 8.1
Kansei quality management.

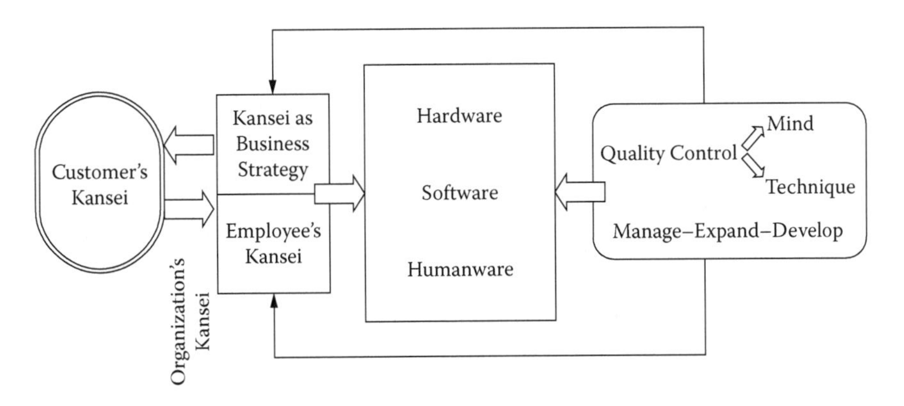

FIGURE 8.2
Structural diagram of Kansei quality management system.

expressed as in Figure 8.2. As shown in the figure, a product that is developed based on the identified Kansei as a business strategy is proposed to a customer. The customer will then verify whether it matches his Kansei requirements. In this sense, since the employee's Kansei also contributes quite an influence, it is matched with the customer's Kansei.

In the underlying background of matching the Kansei of the corporate entity with that of the customer, Kansei such as hardware, software, humanware, and so forth connect and react with one another. Here, the challenge is to be thorough in controls such as the following, as Kansei quality management:

- Grasp the customer's Kansei
- Verify whether the Kansei is correct
- Develop products based on the Kansei
- Train in Kansei to make it possible
- Develop and improve the quality of Kansei service

8.4 Kansei Quality Management Case Studies

8.4.1 Kansei Quality Management for Restaurants

Quality control and company-wide quality control are being introduced in the service industry. The service industry is quite passionate about conducting activities like extracting control items that will fit in the policy. However, it is typical that each individual department, for example, Reception, Kitchen, Service, and Accounting, practice its quality control separately. In Kansei quality management, the policy is to provide a consistent Kansei all the way

FIGURE 8.3
Hospitality Kansei engineering.

through reception, dining (offering of products and services), and payment (Figure 8.3). These are all extracted from the customer's viewpoint and analyzed for consistent quality requirements in response (control) in order to raise the level of customer satisfaction. Here is a real example.

Visualize a customer by the name of Mr. Tanaka entering a restaurant.

- At the reception desk, Mr. Tanaka says, "I'm Tanaka from such and such company. I believe I have reservations for a function at six o'clock." Here, the receptionist grasps Mr. Tanaka's Kansei (the organizer for a gathering of 20 people). Normally, a gathering organizer is the one who selects the venue. By providing good service and treatment that makes the organizer happy, he will come again. Therefore, treating him as an organizer will produce a totally different result than treating him as a normal customer.
- Next, the event is launched. The hot food that was ordered is brought in. This time, the food and cooking style that match the expectations of the Kansei are focused on. When his colleagues comment that they are stuffed and the food was delicious, he will feel that he has successfully performed his responsibility as an organizer.
- Next, when Mr. Tanaka is paying at the counter before leaving, it is important to ask whether the event was as he had expected. Besides ensuring the amount charged is accurate without making him wait, if there are also some words of appreciation, he will be happy.

In ideal quality control, consideration and pursuit of the required quality are done at the respective sections. However, in reality, one person might move through separate sections. Therefore, a consistent service (here, it refers to grasping a consistent quality requirement) is most important. If the separate sections of the restaurant discuss the consistent service to be given to Mr. Tanaka, a Kansei quality management can be accomplished. Mr. Tanaka will get the expected service, and he will surely be very satisfied. Continuing to address him by name could also be an example of the simplest Kansei quality management.

8.4.2 Quality Table for Hotels

Similar to restaurants, hotels also handle three Kansei: the hardware (building), software (marketing system), and humanware (service staff). Let's look at these three Kansei using a hotel as the case example.

Table 8.1 is a segment of a quality table. The following is an example of hardware:

- Kansei quality requirement—Analyze the hotel design from the Kansei viewpoint. The Kansei quality requirement for the hotel is *elegant hotel*. The word *elegant* is the Kansei question. It is part of the Kansei strategy of the company.
- Definition—The required quality is to be seen as elegant hardware from the customer's viewpoint. Therefore, we need to consider the interior design elements that correspond to that. For this, the Kansei engineering techniques that have been described earlier will be useful.
- Content—For the design elements, we need to think of how to elicit an elegant feeling by putting an emphasis on color and color tones. Additionally, we might also need to consider the structural elements such as room size and height.
- Level of importance—Decide on the priority of the design elements that should receive a bigger portion of the investment to elicit the effect.
- Control item—Prepare the control standard and guideline to realize the customer's Kansei quality requirement of *elegant*. In addition to operating according to those standards and guidelines, apply them as well to the Kansei quality requirement in other branches.

For another useful idea, think about *stay in a relaxed hotel* as another Kansei quality requirement (from the viewpoint of the customer), and the Kansei *relaxed*.

This technique is aimed at realizing customer satisfaction. It is an approach in which the Kansei from the customer's viewpoint is grasped from a sepa-

TABLE 8.1

Quality Table for Hotels

		Kansei Quality Requirement	Definition	Content	Level of Importance	Control Item	QCC Relevance
Hardware	① Hotel Design	① Elegant hotel	"Elegant"	① Color tone - Brightness down Saturation down Hue Warm colors - Orange Cold colors - Blue ② Emphasis on hue	○ ○ △ △ △	Construction Design Std. Interior Std. Room Interior Std. Furnishings Design Std. Furnishings Selection Std.	
		⑤ Stay with relaxed-mind hotel	"Relaxed"	① Quite wide room ② Quite wide toilet ③ Don't make windows small ④ Wide bathroom mirror ⑤ Sofa (3-piece)	○ ○ ○ △ ○	Room Area Std. Furnishings Selection	
③ Software	◯ 28 Dealing with customer	◯ 53 Stay comfortably	"Comfortably"	① Family-like interaction Friendly greeting Greet when meet Smile Greet when going out ② Feel as being welcomed Help with the luggage Address by name	○ ○ ○ ○ ○ ○	Kansei-filled Greeting Manual How to Bow Manual Smile Manual Luggage Support Manual	○ ○ ○ ○

rate survey or by paying attention to conversations, set to the required quality in the TQC concept, and enhanced. If the Kansei is considered in the details of the software, hardware, and humanware, we can obtain a more detailed procedure and guidelines than the conventional TQC. This will materialize as a hotel with services that fulfill customer satisfaction. If the software and hardware of this technique are combined with the QC movement and expanded to the Kansei QC movement activities, this might lead to a reformation of service activities.

8.4.3 Kansei at Ladies' Wear Department

Let me explain another simple example. We will change the subject, and I will explain the Kansei quality requirement for a ladies' wear department. This time, I will focus on how to gather and think about the Kansei.

Table 8.2 relates the conventional quality requirements in the left column to the Kansei quality requirements in the right column. The conventional quality deployment in the left column lists the ideal situations as the quality requirements. There are four items listed: *Can do shopping safely and comfortably, Value-for-money products, Can buy desired product at a desired time,* and *Easy to see and easy to take.*

The Kansei quality requirements in the right column are the list of Kansei that consider the mindset of women when they are shopping, with *Can have a pleasant dress-selection.* The concepts that come under it are *Fun to see, Fun to touch, Fun to walk around, Can experience the beauty, Can choose the desired item,* and lastly, *Satisfied to buy.* With these, we can imagine the mindset of women doing their shopping in a joyful mood. This is exactly the idea of customer satisfaction. The quality requirements in the left column seem to be the customer satisfaction considerations, but they are actually from the logic of the companies that provide the products.

The new Q2 table that can be expanded from Table 8.2 is Table 8.3. It is a roughly prepared table, not a completed one.

First of all, for the first-order Kansei, *Can have a pleasant dress-selection,* there is *Fun to see* as the second-order Kansei. Further breakdown of the

TABLE 8.2

Old and New Quality Requirements Mapping for a Ladies' Wear Department

Quality Requirements	Kansei Quality Requirements
Ideal Situations of Ladies' Wear	*Can have a pleasant dress-selection*
1. Facilities and maintenance are in order to enable safe and comfortable shopping	1. Fun to see
1.2 Can do shopping comfortably	2. Fun to touch
1.3 Good environment	3. Fun to walk around
2. Trusted for value-for-money products	4. Can experience the beauty (image of beauty)
2.1 Product level of quality (taste, freshness, etc.) is high	5. Can choose the desired item
2.2 Prices are low	6. Satisfied to buy
3. Can buy desired product at desired time in desired quantities	
3.1 Good varieties of products	
3.2 Can casually buy at desired time in desired quantities	
4. Easy to see and easy to take	
4.1 The sales floor is easy to understand	
4.2 Products can be selected easily	

TABLE 8.3
New Q2 Table

1st-Order		2nd-Order		3rd-Order		Present Situation	Judge	Judgment
1	Can have pleasant dress-selection, pleasant shopping	11	Fun to see	111	Aisles are easy to understand	Aisles are devious	3	Is Layout Manual (LM) available?
				112	Aisles are wide	There are some narrow areas	2	Does LM fit with the times?
				113	Layout is well marked	All sorts with no standard	2	
				114	Product classification by layout is clear	There are areas where product classification is unclear	3	Does LM match Kansei?
				115	Racks are easily viewable			
				116	Arranged for ease in selecting			
				117	Appropriate product quantity in relation to face area			Is Face Manual available? Does Face Manual match Kansei?
				118	Brightly lit			Is there any lighting standard to make dress look nice?
				119	Appropriate illumination for dress selecting			
				120	Mannequins resemble women			Is Mannequin Fitting Manual available?
				121	Mannequin's clothing fits well			Is Product Arrangement Standard available?
				122	Products are properly arranged			

second-order Kansei will be the third-order Kansei. Here, the *Fun to see* Kansei is investigated in the sequence of hardware, software, and humanware, then continues with checking the present situation of each of those items, and further with the judgment for presence or absence of control items. That means we can prepare guidelines as the control items for the judgment items.

As you can see from the table, there is a significant difference between them. As opposed to the conventional Q2 table that has many control items related to hardware, the Kansei quality management has more content related to customer satisfaction and comfort.

8.4.4 Cash Register Work at Supermarket

Finally, I would like to describe Kansei quality management, which is Kansei engineering Type V, as applied to a supermarket. This has been done by the people who are actually involved in cash register work and are cash register instructors, in collaboration with Izumi Co. Ltd., the recipient of the China Quality Management Prize (a prize similar to the Deming Prize).

For this case of Kansei quality management, we have decided to follow these steps.

1. What is the customer's Kansei? We explained to the cash register instructors the details of the work we were going to perform. We especially emphasized that the ultimate goal of the supermarket is customer-oriented sales, and a technique that focuses on customers' Kansei must be adopted.
2. Market-oriented philosophy. We asked them to imagine grasping the customer's Kansei and valuing the customers. We specifically asked them to think about bad examples at the cash register. We asked them to imagine scenarios such as checkout mistakes, making customers wait and thus having them become irritated, and so on, and to think of the true meaning of valuing the customers.
3. Q2 table for cash registers. All of us referred to the conventional quality requirements table (Q2 table as in Table 8.4). We proposed to improve one of the items in the table, the checkout, using a new technique, Kansei quality management.
4. Deliberation of Kansei words. All of us brainstormed the Kansei keywords expected regarding service at the cash register from the customer's viewpoint, and then discussed the meaning of each word (Table 8.4).
5. New Q2 table first-order analysis. While assessing the Kansei keywords obtained in Step 4, we analyzed the Kansei quality requirements related to the cash register, as perceived by customers. Here, we discovered *Kind and pleasant response* Kansei (Table 8.5).

TABLE 8.4

Q2 Table

6. Kind and pleasant response	61 Employee's attitude is good	611	Treat customers with cheerful expression and smile		Employee's expression	Score by monitors
		612	Neat dress and appearance		Employee's appearance	Score by monitors
		613	Polite words and attitude		Employee's discipline	Score by instructors
		614	Careful in handling products		Employee's discipline	Score by instructors
		615	Alert and speedy service		Service level	Score by instructors
	62 Level of service is high	621	Does not reply while continuing work (half-heartedly)		Service level	Customer service contest score
		622	Addressing timing is good		Customer service skill	Customer service contest score
		623	Rich with accurate product knowledge	O	Employee's product knowledge	Customer service contest score
		624	Advise various beneficial information		Employee's information volume	No. of compliments from customers
		625	There's always someone to ask		Number of employees	Customer service MH amount
		626	Floor guide is fast and accurate	O	Employee's floor knowledge	Floor location test score
	63 Response when receiving complaint is good	631	Offer consultation willingly		Customer service skill	Score by monitors
		632	Willingly exchange the returned item		Exchange returned item	Sales: Returned item ratio
		633	During a claim, sincerity is observed without sending customer to numerous people		Complaint handling skill	Score by monitors
		634	No discrimination between people		Service level	Score by monitors
	64 Checkout is fast and accurate	641	Do not keep customer waiting	O	Counter waiting time	No. of waiting customers
		642	No mistakes like wrong change, wrong register, etc.	O	Accurateness of checkout	Number of checkout mistakes
		643	Careful in handling products		Checkout service level	Checker contest score
		644	Service by checker	O	Checkout	Checker contest

New Q2 Table

Casual	Cleanliness	Kindness
Neat	Speed	Gentle
Relaxed	Smile	Patience
Crisp	Richness	Grace
Brisk	Favorable	Careful
Rhythmical	Lovely	Laid-back
Refreshing	Energetic	Organized
Orderly		Animated
Vigorous		Healthy
Smiling		Attentive
		Bright
Happy		Calm
Precisely		
Serene		

Keywords expected from cash register

Summarize Kansei quality requirements from customer's standpoint

6. New Q2 table second-order analysis. We further analyzed the result of the first-order analysis to the second-order Kansei. Here we entered 61: *A checker with a likable personality*; 62: *A checker who can give pleasant service*; and 63: *Not making customers feel irritated at cash register*.
7. New Q2 table third-order analysis. Next, we moved into the third-order analysis. Here, for item 61, we analyzed and entered 611: *Give response cheerfully and lively* through 618: *Goods handling is prompt and careful*; while for item 62, we entered 621 through 628, and for item 63, from 631 through 638. The behavior quality of the checker that is expected by customers has been discovered accurately and in great detail.
8. New Q2 table fourth-order analysis. With the third-order analysis as a clue, we proceed to the fourth-order analysis for more detail. As Table 8.5 shows, from 6111 to 6114, from 6121 to 6123, and so on, the expected behaviors become more comprehensive. Upon reaching this level, we discovered the detailed behaviors, and it became very clear how the cashiers should perform.
9. Assurance item. For the behavior indicator in the fourth-order analysis, we put them in easy-to-understand words as Kansei qualities, and documented clearly the actions to be used as a quality control.
10. Control item. For the action verification items, we checked, evaluated, and then established the control items that will be used as a guide for training to achieve the assurance level, as well as the evaluation systems and manuals.

Table 8.5 shows the differences compared to the conventional Q2 table. We have received favorable comments from the Izumi staff who participated in this analysis, such as:

1. The method of thinking is clear, and it is easy to change past actions.
2. It is very easy to put the new procedures into a manual.
3. Customer satisfaction is easy to understand at the execution level.
4. Evaluation is also easy to understand and easy to justify.

In this chapter, I have described the application of Kansei quality management in the service industry. However, it can also be applied to production shop floors and supporting departments. Since we can standardize the expected behaviors with clues obtained from customers' Kansei, customer-conscious behavior transformation is possible. On top of that, we can achieve customer satisfaction, and at the same time, employee motivation will increase.

TABLE 8.5

Kansei Quality Requirement Extended Table

Kansei Quality Requirement								No. 1	
1st-Order		**2nd-Order**		**3rd-Order**		**4th-Order**		**Assurance Item**	**Control Item**
6	Kind and pleasant response	61	A checker with a likable personality	611	Give response cheerfully and lively	6111	Always respond with a smile	Greet with a smile	Greeting assessment
						6112	Look up and greet	Greet while looking up	Greeting assessment
						6113	Greet with a staccato voice	Staccato voice	Greeting assessment
						6114	Addressing timing is good	Good timing of addressing	Response assessment
				612	Neat dress and appearance	6121	Proper hairstyle that is not hiding the face	Eyes/face can be seen	Appearance assessment
						6122	Uniform/apron is nice without dirt	Nice dress	Appearance assessment
						6123	Name on the nametag is readable	Name is readable	Appearance assessment
						6124	Other appearance is neat	Neat appearance	Appearance assessment
				613	Counter staff who can encourage customers to line up	6131	Respond with a smile to anybody	Always smiling	Greeting assessment
						6132	Looks calm and reassuring	Calm actions	Action assessment
						6133	Seen as taking pride in working	Diplomatic response	Action assessment
						6134	Looks speedy in operating the cash register	Speedy cash register operation	Register operation assessment

614	A checker who makes customers feel like talking	6141	An experienced cashier	Know names and faces	Response assessment
		6142	Rich with knowledge, answer questions pleasantly	Accurate answer	Response assessment
		6143	Always say something to customer	Precise addressing	Greeting assessment
		6144	Make greeting even at sales floor	Greeting at sales floor	Greeting assessment
615	Alert action and speedy	6151	No unnecessary action in operating cash register	Smooth register operation	Register operation assessment
		6152	Think about priority, fast in action	Quick work operation	Action assessment
616	Healthy and assured response	6161	Clear voice with intonation	Very audible voice	Response assessment
		6162	Respond with a smile while looking at customer's face	Greet while looking up	Greeting assessment
		6163	Listen attentively, give back-channel feedback	Good listener	Response assessment
617	Verbiage and attitude gentle and polite	6171	Talk in a manner that befits the customer	Way of talking that befits the customer	Response assessment
		6172	Use honorifics (*sonkeigo, teineigo, kenjyougo*) correctly	Correct verbiage	Response assessment
		6173	Calm and classy in actions	Nice attitude and posture	Response assessment

(continued on next page)

TABLE 8.5 (continued)

Kansei Quality Requirement Extended Table

1st-Order	2nd-Order	3rd-Order	4th-Order	Assurance Item	Control Item
		618 Goods handling is prompt and careful	6181 Handle products according to their nature	Correct product handling	Product handling assessment
			6182 Scan products speedily and rhythmically	Quick scanning	Product handling assessment
			6183 Not put the product against the glass when scanning	Careful product handling	Product handling assessment
	62 A checker who can give a pleasant service	621 Attentive and diplomatic response	6211 Respond with initiative	Initiated response	Response assessment
			6212 Make generous efforts	Responsive until the end	Response assessment
			6213 Smooth with no unnecessary actions	Smooth response	Response assessment
		622 Looks at face when addressing	6221 Talk to customers with smiling face	Respond with smile	Response assessment
		623 Addressing timing is good	6231 Speak in a manner according to TPO	Way of talking according to TPO	Response assessment
			6232 Initiate the conversation	Initiate response	Response assessment

Kansei Quality Requirement								No. 2	
1st-Order		2nd-Order		3rd-Order		4th-Order		Assurance Item	Control Item
6	Kind and pleasant response	62	A checker who can give an pleasant service	624	Correct and rich product knowledge	6241	Properly answer customer's question	Answer clearly	Response assessment
						6242	Has a wide knowledge	Correct sales/checkout knowledge	Product knowledge assessment
				625	Able to guide the sales floor correctly, quickly and accurately	6251	Correctly know the sales floor characteristics	Correct sales floor guide	Sales floor guide assessment
						6252	Able to quickly direct to location asked by customer	Quick and accurate guide	Sales floor guide assessment
				626	Answer clearly when asked	6261	Listen carefully to what customer says	Listen carefully	Response assessment
						6262	Understand and reply correctly to customer	Correct reply	Response assessment
				627	Respond in sincere manner even in complaint handling	6271	Listen properly to customer's explanation	Listen carefully	Complaint handling assessment
						6272	Handle the matter through to the end	Handle to the end	Complaint handling assessment
						6273	Make a sincere apology	Apologize from the heart	Complaint handling assessment

(continued on next page)

TABLE 8.5 (continued)

Kansei Quality Requirement Extended Table

1st-Order	2nd-Order	3rd-Order	4th-Order	Assurance Item	Control Item
		628 Converse with delight, give customer a helping hand	6281 Remember customers' names	Know names	Response assessment
			6282 Recognize customers' faces	Know faces	Response assessment
	63 Not make customers feel irritated at cash register	631 Not make customers wait at cash register	6311 Few people waiting at checkout	Not causing irritation	Keep-waiting assessment
			6312 Bagging table is not crowded	No unattended baskets	Bagging table assessment
		632 Perform register and clearance correctly and speedily	6321 Brisk movements with no unnecessary action	Fast cash register	Register operation assessment
			6322 Movements are rhythmical	Fast and smooth checkout	Register operation assessment
			6323 Register is accurate, no mistakes, can feel assured	Accurate checkout	Register operation assessment
		633 Give change correctly and speedily	6331 Separate bills and coins when giving change for easy receiving	Make returned change easy to be received	Money transfer assessment
			6332 Count bills aloud and quickly	Able to use the top five terms	Response assessment
			6333 Count the bills at a visible location	Count bills visibly	Money transfer assessment
			6334 Place the received cash at a visible location	Place in designated tray	Money transfer assessment

	634	Able to handle products carefully and briskly	6341	Arrange according to products' characteristics, and easy to put in bag	Easy-to-pack product arrangement	Product handling assessment
			6342	Handle products carefully during checkout	Careful product handling	Product handling assessment
			6343	Detect the barcode immediately, position the product at appropriate angle	Fast barcode confirmation	Register operation assessment
	635	Say "I'm sorry to have kept you waiting" sincerely	6351	Look at customer's face when speaking	Look at face when greeting	Greeting assessment
			6352	Put hands together in front and bow	Nod head, hands clasped	Greeting assessment
			6353	Speak with gentle and sincere eyes	Greet with eye contact	Greeting assessment
	636	Respond to cash register mistake in an organized manner	6361	Listen attentively to what customer says	Listen properly to complaint	Complaint handling assessment
			6362	Properly explain the mistake	Correct explanation	Complaint handling assessment
			6363	Quickly fix the problem	Fast posttreatment	Complaint handling assessment
			6364	Properly offer an apology	Can apologize properly	Complaint handling assessment

(continued on next page)

TABLE 8.5 (continued)

Kansei Quality Requirement Extended Table

1st-Order	2nd-Order	3rd-Order		4th-Order		Assurance Item	Control Item
		637	Treat customers fairly	6371	Give the same treatment to any customer	Fair response	Response assessment
				6372	Handle products in the same way according to their nature for every customer	Fair product handling	Product handling assessment
				6373	Properly pack in the product accessory for every customer	Fair product handling	Product handling assessment
		638	Response in quick-witted manner	6381	Offer help when customer in trouble	Offer help	Response assessment
				6382	Be the first one to show concern to customers	Offer help	Response assessment
				6383	Take appropriate action for any accident, and follow through until settled	Quick-witted response	Response assessment
				6384	Respond to a forwarding in a smooth, prompt, and assured manner	Correct forwarding	Forwarding assessment

Index

A

Adjective pairs, 41
Articulation
  anatomical chart, 38
  categories, 39–40
Artificial intelligence system, 80–83
  prototyping, 80

B

Backward Kansei engineering, 98–99, 102–103
  with template, 100–105
Brain wave measurement, 52
Brassiere development, Kansei engineering, 13–15, 30–31

C

Cars, Kansei engineering, 11–13, 92–95
Case studies of Kansei engineering, 11–22
  brassieres, 13–15, 30–31
  cars, 11–13, 92–95
  evaluation experiments, 16–17
  grievances, causes of, 15
  housing, 15–19
  Kansei words, collecting, 16
  word sound image, 19–21
Cash register work, total quality control, 125–134
Causes of grievances, 15
Causes of human error, 112
Cleanliness, inference technique for, 103
Client interface, artificial intelligence system, 82
Cognition, as source of human error, 112
Communication, 27–28
Comprehensive nature of Kansei, 5–7
Computer-generated virtual world, 107–108. *See also* Virtual reality
Computer hardware, virtual reality, 107
Computer memory content, 109
Configuration of semantic differential scale, 61–63

D

Database, 78–79
  design elements database, 78–79
  Kansei words database, 78
  rule base, 79
Decision-making by consumer, 36–37, 111–112
Definition of Kansei, 4–5
Design elements, 32
Design requirement, 32
Designer support system, 36–37
Developing valuable product, 1–2
Development, sales in Kansei, 8–10
Diagnostic system, word sound image, 44–47
Direct feeling, 27–29
Door, type II Kansei engineering, 89–91
Driver wish, 28

E

ECG. *See* Electrocardiogram
Education, 112–113
Electrocardiogram, 52
Electromyography, 52
Emergence of Kansei engineering, 1–10
EMG. *See* Electromyography
Emotion, as source of human error, 112
Engineering case studies
  brassieres, 13–15, 30–31
  cars, 11–13, 92–95
  evaluation experiments, 16–17
  grievances, causes of, 15
  housing, 15–19
  Kansei words, collecting, 16
  word sound image, 19–21

Engineering procedures
  artificial intelligence system, 80–83
  brain wave measurement, 52
  database, 78–79
    design elements database, 78–79
    Kansei words database, 78
    rule base, 79
  electrocardiogram, 52
  electromyography, 52
  ergonomic indicators, physiological, 52
  evaluation sample preparation, 64–65
  high-level Kansei word extraction, 59–61
  item/category extraction, 65–67
  Kansei words factor analysis, 57–59
  low-level Kansei words, extracting, 52–53
  physiological measurement, 52
  primary evaluation experiment, 55–57
  sample evaluation sheet, 62
  secondary evaluation experiment, 63–64
  semantic differential scale configuration, 61–63
  semantic differential scale construction, 54–55
  spacious feeling, 66
  statistical analysis, 68–78
    Kansei word factor analysis, 68
    multivariate analysis, 69–70
    quantification theory type I, 70–78
    semantic differential evaluation analysis, 68–69
  survey target selection, 49–51
  type II Kansei engineering, 31–37, 49–96
Entrance door, type II Kansei engineering, 89–91
Epiglottis, 38
Ergonomic indicators, physiological, 52
Error, human
  causes of, 112
  cognition as source of, 112
  emotion as source of, 112
  risk of, 112
Esophagus, 38
Evaluation experiments, 16–17
Evaluation sample preparation, 64–65

F

Fashion image system, type II Kansei engineering, 88–89
Forward Kansei engineering, 97–98
  consumer product selection, 97
  designer product development, 97–98

G

Good product, defined, 7
Graphic display, artificial intelligence system, 81
Grievances, causes of, 15

H

Hard palate, 38
Head-mounted display, virtual reality, 108
High-level Kansei word extraction, 59–61
HMD. *See* Head-mounted display
Hotels, total quality control, 121–122
Housing, Kansei engineering, 15–19
HULIS. *See* Human living system
Human error
  causes of, 112
  cognition as source of, 112
  emotion as source of, 112
  risk of, 112
Human living system, type II Kansei engineering, 85–88
Hybrid Kansei engineering, 97–106
  backward Kansei engineering, 98–99, 102–103
    with template, 100–105
  cleanliness, inference technique for, 103
  forward Kansei engineering, 97–98
    consumer product selection, 97
    designer product development, 97–98
  future of, 104–105
  image recognition procedure, 101–102
  openness, inference technique for, 103
  recognition for input image information, 101–102

recognition results example, 103–104
unconstrained, inference technique for, 103

I

Image recognition procedure, 101–102
Interference processor, artificial intelligence system, 80–81
Item/category extraction, 65–67

J

Japanese vowel categories, 40

K

Kansei engineering
backward, 98–105
brassieres, 13–15, 30–31
cars, 11–13, 92–95
case studies, 11–22
comprehensive nature of, 5–7
custom kitchen, 109–110
definition of, 4–5
elements, 6
emergence of, 2–4
extracting high-level words, 59–61
extracting low-level words, 52–53
factor analysis of words, 57–59
forward, 97–99
housing, 15–19
hybrid, 97–106
ladies' wear department, 123–125
overview, 1–10
physical characteristics translation map, 29
procedures, 49–84
quality control, 115–134
quality management, 115–134
case studies, 119–134
cash register work, supermarket, 125–134
defined, 115–116
for hotels, 121–122
at ladies' wear department, 123–125
for restaurants, 119–121
supermarket cash register work, 125–134
techniques, 23–48
total quality control, 115–134
type I, 23–31
type II, 31–37, 49–96
type III, 37–47
type IV, 107–114
virtual, 107–114
zero-order concept, 27–28
Kansei words
collecting, 16
database, 78
evaluation of pencils using, 72
evaluation sheet, 62
factor analysis, 57–59, 68
factor analysis of, 68
high-level, extracting, 59–61
low-level, extracting, 52–53
sample evaluation sheet, 62
semantic differential, 38–44
survey on, 38–44
sound image, 19–21
diagnostic system, 44–47

L

Ladies' wear department, total quality control, 123–125
Larynx, 38
Low-level Kansei words, extracting, 52–53
Lower lip, 38

M

Maneuvering feeling, 28
Manners of articulation categories, 40
Mouth, 38

N

Narrowness, appropriate, 28
Naval cavity, 38
Nose, 38
Nostril, 38

O

Openness, inference technique for, 103

P

Passenger car design application, type I Kansei engineering, 27–30
Pharynx, 38
Phonetic symbols, 39
Physical design characteristics deployment, type I Kansei engineering, 26–27
Physiological measurement, 52
Places of articulation categories, 39
Primary evaluation experiment, 55–57
Product concept breakdown, type I Kansei engineering, 24–26
Product concept determination, type I Kansei engineering, 23–24

Q

Quality management, 115–134
  case studies, 119–134
  for hotels, 121–122
  at ladies' wear department, 123–125
  for restaurants, 119–121
  supermarket cash register work, 125–134

R

Recognition for input image information, 101–102
Recognition results example, 103–104
Restaurants, total quality control, 119–121
Risk of human error, 112
Running feeling, 27–29, 33

S

Sales, development in Kansei, 8–10
Sample evaluation sheet, 62
Secondary evaluation experiment, 63–64
Self-control, 32–33
Semantic differential of words, 38–44
Semantic differential scale
  configuration, 61–63
  construction, 54–55
Soft engineering. *See* Kansei engineering
Soft palate, 38
Sound, hardware generating, 37–38
Spacious feeling, 66
Statistical analysis, 68–78
  Kansei word factor analysis, 68
  multivariate analysis, 69–70
  quantification theory type I, 70–78
  semantic differential evaluation analysis, 68–69
Statistical processor flowchart, 95
Style selection, 109–110
Supermarket cash register work, total quality control, 125–134
Survey target selection, 49–51
System control, artificial intelligence system, 82

T

Target identification, type I Kansei engineering, 23
Technical specifications, type I Kansei engineering, 27
Techniques in Kansei engineering, 23–48
Tight feeling, 27–29
Tongue, 38
Total quality control, 115–134
  case studies, 119–134
  for hotels, 121–122
  at ladies' wear department, 123–125
  for restaurants, 119–121
  supermarket cash register work, 125–134
TQC. *See* Total quality control
Trachea, 38
Training, 112–113
Translation technique, 32
Type I Kansei engineering, 23–31
  brassiere development application, 13–15, 30–31
  overview, 23–27
  passenger car design application, 27–30
  physical design characteristics deployment, 26–27
  product concept breakdown, 24–26
  product concept determination, 23–24

target identification, 23
technical specifications, 27
zero-order Kansei concept, 27–28
Type II Kansei engineering, 31–37, 49–96
application, 34–36
artificial intelligence system, 80–83
brain wave measurement, 52
car interior, 92–95
cases, 85–96
consumer decision-making system, 36–37
database, 78–79
design elements database, 78–79
Kansei words database, 78
rule base, 79
design element, 32
design requirement, 32
designer support system, 36–37
electrocardiogram, 52
electromyography, 52
entrance door, 89–91
ergonomic indicators, physiological, 52
evaluation sample preparation, 64–65
fashion image system, 88–89
high-level Kansei word extraction, 59–61
human living system, 85–88
elements, 85
item/category extraction, 65–67
Kansei words factor analysis, 57–59
low-level Kansei words, extracting, 52–53
overview, 31–34
physiological measurement, 52
primary evaluation experiment, 55–57
running feeling, 29, 33
sample evaluation sheet, 62
secondary evaluation experiment, 63–64
self-control, 32–33
semantic differential scale configuration, 61–63
semantic differential scale construction, 54–55
spacious feeling, 66
statistical analysis, 68–78
Kansei word factor analysis, 68
multivariate analysis, 69–70
quantification theory type I, 70–78
semantic differential evaluation analysis, 68–69
statistical processor flowchart, 95
survey target selection, 49–51
translation technique, 32
Type III Kansei engineering, 37–47
articulation, anatomical chart, 38
semantic differential of words, 38–44
sound, hardware generating, 37–38
word sound image diagnostic system, 44–47
Type IV Kansei engineering, 107–114
causes of human error, 112
cognition, as source of human error, 112
computer memory content, 109
custom kitchen Kansei engineering, 109–110
customer decision-making, 111–112
education, 112–113
emotion, as source of human error, 112
evolution, 110–113
risk of human error, 112
style selection, 109–110
training, 112–113
virtual reality, 107–108, 110
computer hardware, 107
head-mounted display, 108

U

Unconstrained, inference technique for, 103
Upper lip, 38
Uvula, 38

V

Valuable product, developing, 1–2
Virtual Kansei engineering, 107–114
causes of human error, 112
cognition, as source of human error, 112
computer memory content, 109
custom kitchen Kansei engineering, 109–110

customer decision-making, 111–112
education, 112–113
emotion, as source of human error, 112
evolution, 110–113
risk of human error, 112
style selection, 109–110
training, 112–113
virtual reality, 107–108, 110
    computer hardware, 107
    head-mounted display, 108
Virtual reality, 107–108, 110
    computer hardware, 107
    head-mounted display, 108
Vocal chord, 38

W

Words, Kansei. *See* Kansei words

Z

Zero-order concept, 27–28